THE FUNDAMENTAL PRINCIPLE OF DIGITS OF A NUMBER
d = 9 k

CHIBAMBA MULENGA, PH.D.

outskirts
press

The Fundamental Principle of Digits of a Number
d = 9 k
All Rights Reserved.
Copyright © 2020 Chibamba Mulenga, PH.D.
v2.0 R2

The opinions expressed in this book are solely the opinions of the author and do not represent the opinions or thoughts of the publisher. The author has represented and warranted full ownership and/or legal right to publish all the materials in this book.

This book may not be reproduced, transmitted, or stored in whole or in part by any means, including graphic, electronic, or mechanical without the express written consent of the publisher except in the case of brief quotations embodied in critical articles and reviews.

Outskirts Press, Inc.
http://www.outskirtspress.com

ISBN: 978-1-9772-3144-4

Library of Congress Control Number: 2020913744

Infinity Symbol Cover Image © 2020 Victor De Schwanberg/Science Photo Library/Science Photo Library via Getty Images. All rights reserved - used with permission.

Outskirts Press and the "OP" logo are trademarks belonging to Outskirts Press, Inc.

PRINTED IN THE UNITED STATES OF AMERICA

DEDICATION

In loving memory of my mother, Mwelwa Mitalabi; my uncle Chewe Mitalabi; and my grandmother, Kafi Nsankinta, a storyteller par excellence of Bemba royal lineage … and to wayfarers in the ocean of life who rendered me kindness and support.

ACKNOWLEDGMENTS

I render gratitude to my Senior publishing consultant and to the Outskirts Press staff, for making the publication of this book possible.

PREFACE

The purpose of this book is to introduce a new mathematical principle. The mathematical principle is related to the rule of divisibility of a number by 9. But the new mathematical principle does not depend on this rule. We refer to the new mathematical principle as The Fundamental Principle of Digits of a Number.

The rule of divisibility of a number by 9 is generally found in books which are concerned with basic math and pre-algebra.

The Fundamental Principle of Digits of a Number is a new mathematical idea for which I received a copyright from the United States Library of Congress in 2014. This new mathematical idea occurred to me as follows.

One summer afternoon in the Tenderloin District of San Francisco, California, I was resting on my bed looking at the ceiling above me, but with no particular concern on my mind.

Suddenly, I saw the number 12 appear on my inner eye. The number 12 turned around and formed the number 21. The two numbers subtracted, and the number 9 appeared. Then the three numbers quickly disappeared from my inner eye.

This event seemed so miraculous to me that I immediately sat up to try and digest it. For I had never before seen a number move by its own power as if it were a bird in the sky.

I was so moved by this event that I resolved to investigate it with the digits of other numbers as far as possible. I did not carry out this investigation all at once, but over a period of time.

In the early stages of this investigation, I proceeded as follows. In each case I picked a number, formed a second numerical quantity by arranging the digits of the given number in reverse order, and determined the difference between the two numerical quantities. I referred to the three steps as the reverse order procedure.

First, I applied the reverse order procedure to 2-digit numbers and made these observations. If the procedure was applied to a given number in which the digit-difference, the difference between the two digits of the given number, was 1, then the difference determined in the procedure was 9. For example, applying the procedure to 43 yields the difference as 43 − 34 = 9. Likewise, applying the procedure to 12 yields the difference as 21 − 12 = 9, just like it appeared in this event. If the procedure was applied to a given number in which the digit-difference was 2, then the difference determined in the procedure was 18, and so on. Thus the difference which the reverse order procedure yielded in each case could be expressed as a product of the digit-difference and 9, the digit-difference-multiplier for any 2-digit number whose digits are distinct.

Second, I applied the procedure to 3-digit numbers and made these observations. If the procedure was applied to a given number in which the digit-difference, the difference between the first and the last digits of a given number, was 1, then the difference determined in the procedure was 99. For example, applying the procedure to 283 yields the difference in the procedure as 382 − 283 = 99. If the procedure was applied to a given number in which the digit-difference was 2, then the difference determined in the procedure was 198, and so on. Thus the difference which the reverse order procedure yielded in each case could be expressed as a product of the digit-difference and 99, the digit-difference-multiplier for any 3-digit number whose digits are distinct, and for any 3-digit modal number as defined in Chapter Two.

Third, I applied the reverse order procedure to 4-digit numbers. In each case the difference determined in the procedure was divisible by 9. But I soon realized that the generalizations similar to the ones that I made if the procedure was applied to 2-digit and 3-digit numbers, could not be made if the procedure was applied to numbers which consisted of more than three digits.

However, similar generalizations could be made if the reverse order procedure was applied to modal numbers which consisted of more than three digits. Thus if the reverse order procedure was applied to a modal number, then the difference determined in the procedure could be expressed as a product of the digit-difference and the corresponding digit-difference-multiplier. For example, applying the procedure to the 4-digit modal number 3001 yields the difference as 3001 − 1003 = 1998 = 2 × 999, where 999 is the digit-difference-multiplier for any 4-digit modal number. Likewise, applying the procedure to the 5-digit modal number 42227 yields the difference as 72224 - 42227 = 29997 = 3 × 9999, where 9999 is the digit-difference-multiplier for any 5-digit modal number, and so on.

Several weeks after I started this investigation, a related new idea unexpectedly appeared in my mind. I picked a 4-digit number in which not all digits were the same. First, I formed a new numerical quantity from that number by arranging the digits of the given number in any order such that this numerical quantity had the same digits and the same number of digits as the given number. I referred to any numerical quantity that I formed in this way as a permutation of digits of the given number. Second, I formed another permutation of digits of the given number in the same way. Finally, I subtracted the two permutations of digits of the given number, and the difference between them was divisible by 9. This was an amazing result.

This result took my investigation of digits of a number to the next level. I referred to the steps involved in determining the difference between two permutations of digits of a given number as the fundamental procedure.

I applied the fundamental procedure to as many numbers as I could up to 15-digit numbers in which not all digits were the same. In each case the difference determined in the procedure was divisible by 9. In some cases it was clear at a glance that the difference was divisible by 9. Otherwise, I used the rule of divisibility of number by 9 to check that the difference determined in the procedure was divisible by 9. For example, two of the permutations of digits of the number 317 are 731 and 137. The difference between 731 and 137 is 731 − 137 = 594. The sum of the digits of 594 is 18, which is divisible by 9. Since the sum of the digits of 594 is divisible by 9, then by the rule of divisibility of a number by 9, the number 594 is divisible by 9. That is all there is to the rule of divisibility of a number by 9.

The results which I obtained by applying the fundamental procedure to different numbers in which digits were distinct, convinced me that there was a mathematical law that governed the difference between two permutations of digits of a given number. The results also confirmed that the reverse order procedure was a special case of the fundamental procedure.

As my personal circumstances directed me to do so, I left San Francisco and moved to Seattle, Washington.

It was in Seattle that I formulated the mathematical law that governed the difference between two permutations of digits of a given number. I referred to this mathematical law as The Fundamental Principle of Digits of a Number.

Stating The Fundamental Principle of Digits of a Number was a major breakthrough in my investigation of digits of a number.

The importance of The Fundamental Principle of Digits of a Number in this investigation cannot be underestimated. For all new concepts, discoveries, and results on digits of a number depended on this principle.

Subsequently, I focused my attention on inventing number patterns and applying the reverse order procedure to them to generate infinite sequences of multiples of 9. Determining the nth terms of these infinite sequences of multiples of 9 became almost like a pastime.

I named infinite number patterns as primary, subprimary, secondary, and subsecondary number patterns. I named infinite sequences of multiples of 9 as primary, subprimary, secondary and subsecondary sequences. I defined the infinite number patterns and infinite sequences as follows.

A primary number pattern is an infinite number pattern of the form

$$a, ac, abc, abbc, abbbc \ldots$$

where a, b, and c are digits, a ≠ 0, and the absolute value of the difference between a and c is 1.

A primary sequence is an infinite sequence of multiples of 9 that is formed if the reverse order procedure is applied to a primary number pattern and the given multiples of 9 are arranged in ascending order.

A subprimary number pattern is an infinite number pattern whose terms are determined by its natural order of terms, an infinite arithmetic sequence of positive integers with specified first term and common difference such that its terms indicate which terms of a given primary number pattern in their sequential order, are terms of a subprimary number pattern.

A subprimary sequence is an infinite sequence of multiples of 9 that is formed if the reverse order procedure is applied to a subprimary number pattern and the given multiples of 9 are arranged in ascending order.

A secondary number pattern is an infinite number pattern of the form

$$ab, cabc, ccabcc, cccabccc \ldots$$

where a, b, and c are digits, a ≠ 0, c ≠ 0, and the absolute value of the difference between a and b is 1.

A secondary sequence is an infinite sequence of multiples of 9 that is formed if the reverse order procedure is applied to a secondary number pattern and the given multiples of 9 are arranged in ascending order.

A subsecondary number pattern is an infinite number pattern whose terms are determined by its natural order of terms, an infinite arithmetic sequence of positive integers with specified first term and common difference such that its terms indicate which terms of a given secondary number pattern in their sequential order, are terms of a subsecondary number pattern.

A subsecondary sequence is an infinite sequence of multiples of 9 that is formed if the reverse order procedure is applied to a subsecondary number pattern and the given multiples of 9 are arranged in ascending order.

I said that the Fundamental Principle of Digits of a Number is related to the rule of divisibility of a number by 9. For this reason, several books which refer to this rule are included in this treatise as Related Books. However, the related mathematical literature does not mention the Fundamental Principle of Digits of a Number.

This book introduces new concepts, discoveries, and results concerning digits of a number. The book also offers everyone something new about digits of a number which they can learn and investigate on their own.

But all these ideas are based on one mathematical idea, The Fundamental Principle of Digits of a Number.

This mathematical idea, of a completely new and unique type in the mathematical firmament, will enamor, surprise, and perplex the worthy inquirer throughout the corridors of time.

To the interested reader who feels that more knowledge of this new mathematical idea is needed, we present The Fundamental Principle of Digits of a Number.

Chibamba Mulenga

CONTENTS

Chapter 1: The Fundamental Principle of Digits of a Number 1

Chapter 2: Predetermining the Difference Given
in the Reverse Order Procedure 7

Chapter 3: Primary Number Patterns and Primary Sequences 14

Chapter 4: Subprimary Number Patterns and
Subprimary Sequences 24

Chapter 5: Secondary Number Patterns and Secondary Sequences 56

Chapter 6: Subsecondary Number Patterns and
Subsecondary Sequences 67

CHAPTER ONE

The Fundamental Principle of Digits of a Number

The purpose of this Chapter is to introduce a new mathematical principle in the decimal number system. We call the new mathematical principle The Fundamental Principle of Digits of a Number. This mathematical principle is related to a well-known rule of divisibility of a number by 9. The rule states that a number is divisible by 9 if the sum of its digits is divisible by 9. But before presenting the new mathematical principle, we indicate the related terminology.

Definition 1: An integer is any of the elements in the set {. . .-3, -2, -1, 0, 1, 2 . . .}.

Definition 2: A digit in the decimal number system is any of the integers 0, 1, 2, 3, 4, 5, 6, 7, 8, and 9.

Definition 3: A number is any of the integers 0, 1, 2 . . .

Definition 4: A permutation of digits of a given number is an arrangement of the digits of the given number in any order such that the numerical quantity which results from the arrangement of the digits of the given number, has the same digits and the same number of digits as the given number.

Definition 5: The difference between two permutations of digits of a given number is the number that is determined by subtracting the two permutations of digits, subtracting the smaller from the larger permutation of digits if the two permutations of digits are different.

We now state The Fundamental Principle of Digits of a Number as follows.

The Fundamental Principle of Digits of a Number: The difference d between two permutations of digits of a number occurs as a multiple of 9 given by

$$d = 9k$$

for $k = 0, 1, 2 \ldots$

It follows from the preceding equation that the difference between two permutations of digits of a number occurs as one of the numbers

$$0, 9, 18 \ldots$$

Since the difference between two permutations of digits of a number is a multiple of 9, then it is divisible by 9. Accordingly, we can use the rule of divisibility of a number by 9 to check that the difference between two permutations of digits of a given number is divisible by 9.

EXAMPLE 1: Two of the permutations of digits of the number 9040 are 4009 and 0094. The difference between 4009 and 0094 is

$$4009 - 0094 = 3915,$$

which is divisible by 9.

EXAMPLE 2: Two of the permutations of digits of the number 2817 are 7218 and 2817. The difference between 7218 and 2817 is

$$7218 - 2817 = 4401,$$

which is divisible by 9.

EXAMPLE 3: Two of the permutations of digits of the number 54097 are 70945 and 47509. The difference between 70945 and 47509 is

$$70945 - 47509 = 23436,$$

which is divisible by 9.

EXAMPLE 4: Two of the permutations of digits of the number 261843756 are 645628371 and 387461526. The difference between 645628371 and 387461526 is

$$645628371 - 387461526 = 258166845,$$

which is divisible by 9.

Discovery 1: The only two permutations of digits of a number whose digits are identical, are one and the same number.

EXAMPLE 5: The only two permutations of digits of the number 111 are one and the same number 111. The difference between 111 and 111 is

$$111 - 111 = 0.$$

Discovery 2: The only two permutations of the digit of a number that consists of one digit, is one and the same number.

EXAMPLE 6: The only two permutations of the digit 0 of the number 0, are one and the same number 0. The difference between 0 and 0 is

$$0 - 0 = 0.$$

We refer to the infinite sequence 0, 9, 18 . . . as the **fundamental sequence on digits of a number.**

The fundamental sequence on digits of a number is an infinite arithmetic sequence with first term 0, common difference 9, and n^{th} term

$$9(n-1)$$

for n = 1, 2 . . .

Formally, we determine the difference between two permutations of digits of a given number in three steps which we refer to as the **Fundamental Procedure:**

1. Pick a number, and consider the given number itself as a permutation of digits which constitute the number.
2. Form two permutations of digits of the given number, with the possibility that the given number itself could be one of the two permutations of digits of the given number which are formed.
3. Determine the difference between the two permutations of digits of the given number, subtracting the smaller from the larger of the two permutations of digits if the two permutations of digits are different.

Each of the preceding examples involves the use of the fundamental procedure.

A second computing procedure is a special case of the fundamental procedure. The second procedure only determines the difference between a given number and a numerical quantity that results from arranging the digits of a given number in reverse order. We refer to this second procedure as the **Reverse Order Procedure:**

1. Pick a number, and consider the given number as the first numerical quantity which is formed in the procedure.
2. Form a second numerical quantity in the procedure by arranging the digits of the given number in reverse order.
3. Determine the difference between the two numerical quantities which are formed in the procedure, subtracting the smaller from the larger numerical quantity if the two numerical quantities are different.

As in the Fundamental Procedure, the difference which is determined in the Reverse Order Procedure is divisible by 9. It's so because the two numerical quantities which are formed in the Reverse Order Procedure are both permutations of digits of a given number. Hence by The Fundamental Principle of Digits of a Number, the difference between the two numerical quantities which are formed in the Reverse Order Procedure is a multiple of 9.

EXAMPLE 7: Applying the reverse order procedure to 327 gives the difference as

$$723 - 327 = 396,$$

which is divisible by 9.

EXAMPLE 8: Applying the reverse order procedure to 8000 gives the difference as

$$8000 - 0008 = 7992,$$

which is divisible by 9.

EXAMPLE 9: Applying the reverse order procedure to 444 gives the difference as

$$444 - 444 = 0,$$

which is divisible by 9.

EXAMPLE 10: Applying the reverse order procedure to 5013728 gives the difference as

$$8273105 - 5013728 = 3259377,$$

which is divisible by 9.

As a rule the two computing procedures are applied to a number, but not to a numerical quantity which is not a number by the given definition of a number. Thus both procedures can be applied to the number 100 but cannot be applied to the numerical quantity 001, because 001 is not a number by the given definition of a number.

Discovery 3: The difference which occurs if the Reverse Order Procedure is applied to a given number whose digits are symmetric about its middle digit is 0.

EXAMPLE 11: The digits of the number 131 are symmetric about the digit 3. Applying the reverse order procedure to 131 gives the difference as

$$131 - 131 = 0.$$

EXAMPLE 12: The digits of the number 6215126 are symmetric about the digit 5. Applying the reverse order procedure to 6215126 gives the difference as

$$6215126 - 6215126 = 0.$$

EXAMPLE 13: The digits of the number 4998994 are symmetric about the digit 8. Applying the reverse order procedure to 4998994 gives the difference as

$$4998994 - 4998994 = 0.$$

CHAPTER TWO

Predetermining the Difference Given in the Reverse Order Procedure

The purpose of this Chapter is to present formulas that can predetermine the difference which would occur if the reverse order procedure is applied to a given k-digit number to which a formula applies.

Definition 1: A k-digit number is a number that consists of k digits, for k = 1, 2 . . .

Among the types of numbers to which the formulas do not apply are k-digit numbers whose digits are such that if the reverse order procedure is applied to a given number, the difference which would occur in the procedure is 0.

The given formulas apply to three types of numbers: 2-digit numbers in which digits are distinct; 3-digit numbers in which digits are distinct; and modal numbers.

In each case the idea behind a formula is that for the type of numbers to which it applies, a formula can use the difference between the first and the last digits of a given number to predetermine the difference which would occur if the reverse order procedure is applied to the given number.

In what follows we state definitions, discoveries, and formulas which relate to the topic of this Chapter, and are based on insights gained from applying the reverse order procedure to many k-digit numbers. We also give examples to highlight these ideas.

Case 1: 2-digit numbers in which digits are distinct

Definition 2: The difference which occurs if the reverse order procedure is applied to a given 2-digit number whose digits are distinct, is the number that is determined by subtracting the smaller from the larger quantity between the given 2-digit number and the numerical quantity that results from arranging the digits of the given 2-digit number in reverse order.

Definition 3: The digit-difference of a given 2-digit number whose digits are distinct is the number that is determined by subtracting the smaller from the larger digit of the given 2-digit number.

Definition 4: The digit-difference-multiplier of a given 2-digit number whose digits are distinct, is a multiple of 9 by which the digit-difference of the given 2-digit number is multiplied to form a product that is equal to the difference which occurs if the reverse order procedure is applied to the given 2-digit number.

Discovery 1: The digit-difference-multiplier of a given 2-digit number whose digits are distinct is 9.

Discovery 2: Let u be the difference which occurs if the reverse order procedure is applied to a given 2-digit number whose digits are distinct, j the digit-difference, and w the digit-difference-multiplier of the given 2-digit number. Then

$$u = j \times w.$$

EXAMPLE 1: Suppose the given number is 28. Then $j = 6$, $w = 9$, and $j \times w = 6 \times 9 = 54$. Applying the reverse order procedure to 28 gives the difference as $u = 82 - 28 = 54 = j \times w$. Hence the Formula holds.

EXAMLE 2: Suppose the given number is 70. Then j = 7, w = 9, and j × w = 7 × 9 = 63. Applying the reverse order procedure to 70 gives the difference as u = 70 − 07 = 63 = j × w. Hence the Formula holds.

Case 2: 3-digit numbers in which digits are distinct

Definition 5: The difference which occurs if the reverse order procedure is applied to a given 3-digit number whose digits are distinct, is the number that is determined by subtracting the smaller from the larger quantity between the given 3-digit number and the numerical quantity that results from arranging the digits of the given 3-digit number in reverse order.

Definition 6: The digit-difference of a given 3-digit number whose digits are distinct is the number that is determined by subtracting the smaller from the larger quantity between the first and the last digits of the given 3-digit number.

Definition 7: The digit-difference-multiplier of a given 3-digit number whose digits are distinct, is a multiple of 9 by which the digit-difference of the given 3-digit number is multiplied to form a product that is equal to the difference which occurs if the reverse order procedure is applied to the given 3-digit number.

Discovery 3: The digit-difference-multiplier of a given 3-digit number whose digits are distinct is 99.

Discovery 4: Let v be the difference which occurs if the reverse order procedure is applied to a given 3-digit number whose digits are distinct, y the digit-difference, and p the digit-difference-multiplier of the given 3-digit number. Then

$$v = y \times p.$$

EXAMPLE 3: Suppose the given number is 407. Then y = 3, p = 99, and y × p = 3 × 99 = 297. Applying the reverse order procedure to 407 gives the difference as v = 704 − 407 = 297 = y × p. Hence the Formula holds.

EXAMPLE 4: Suppose the given number is 379. Then $y = 6$, $p = 99$, and $y \times p = 6 \times 99 = 594$. Applying the reverse order procedure to 379 gives the difference as $v = 973 - 379 = 574 = y \times p$. Hence the Formula holds.

Case 3: Modal numbers

Definition 8: A modal number is a 3-digit number of the form

$$aac$$

or, it's a 3-digit number of the form

$$acc$$

and, for $k = 4, 5 \ldots$, it's a k-digit number of the form

$$abb \ldots c$$

where a, b, and c are digits, $a \neq 0$, and $a \neq c$.

In particular, if $b = a$ in Definition 8, then $abb \ldots c$ becomes

$$aaa \ldots c$$

and, if $b = c$ in Definition 8, then $abb \ldots c$ becomes

$$acc \ldots c.$$

Definition 9: A k-digit modal number is a modal number that consists of k digits, for $k = 3, 4 \ldots$

EXAMPLE 5: Four of the 3-digit modal numbers are 447, 552, 388, and 100.

EXAMLE 6: Four of the 5-digit modal numbers are 80001, 27774, 59999, and 10000.

Definition 10: The difference which occurs if the reverse order procedure is applied to a given k-digit modal number, is the number that is determined by subtracting the smaller from the larger quantity between the given k-digit modal number and the numerical quantity that results from arranging the digits of the given k-digit modal number in reverse order.

Definition 11: The digit-difference of a given k-digit modal number is the number that is determined by subtracting the smaller from the larger quantity between the first and the last digits of the given k-digit modal number.

Definition 12: The digit-difference-multiplier of a given k-digit modal number is a multiple of 9 by which the digit-difference of the given k-digit modal number is multiplied to form a product that is equal to the difference which occurs if the reverse order procedure is applied to the given k-digit modal number.

Discovery 5: Let q, c and m be functions, k the number of digits in a given modal number, q(k) the difference which occurs if the reverse order procedure is applied to a given modal number, c(k) the digit-difference, and m(k) the digit-difference-multiplier of a given modal number. Then

$$q(k) = c(k) \times m(k)$$

for k = 3, 4 . . .

Discovery 6: The digit-difference-multiplier of a given k-digit modal number is

$$m(k) = 10^{k-1} - 1$$

for k = 3, 4 . . .

It follows from Discovery 6 that the digit-difference-multiplier of a 3-digit modal number is 99.

By Discovery 3 the digit-difference-multiplier of a 3-digit number whose digits are distinct is also 99.

EXAMPLE 7: Suppose the given modal number is 288. Then k = 3, c(3) = 6, m(3) = 99, and c(3) × m(3) = 6 × 99 = 594. Applying the reverse order procedure to 288 gives the difference as q(3) = 882 − 288 = 594 = c(3) × m(3). Hence the Formula holds.

EXAMPLE 8: Suppose the given modal number is 7000. Then k = 4, c(4) = 7, m(4) = 999, and c(4) × m(4) = 7 × 999 = 6993. Applying the reverse order procedure to 7000 gives the difference as q(4) = 7000 − 0007 = 6993 = c(4) × m(4). Hence the Formula holds.

EXAMPLE 9: Suppose the given modal number is 16664. Then k = 5, c(5) = 3, m(5) = 9999, and c(5) × m(5) = 3 × 9999 = 29997. Applying the reverse order procedure to 16664 gives the difference as q(5) = 46661 − 16664 = 29997 = c(5) × m(5). Hence the Formula holds.

We now present a discovery that relates size of a number, digit-difference, and the difference which occurs in the reverse order procedure.

Definition 13: The size of a given number is the number of digits in the given number.

Discovery 7: Two or more given numbers that have the same size and the same digit-difference yield the same difference if the reverse order procedure is applied to each of them.

EXAMPLE 10: The numbers 146 and 449 have the same size of 3 and the same digit-difference of 5. Applying the reverse order procedure to 146 gives the difference as 641 − 146 = 495. Applying the reverse order procedure to 449 gives the difference as 944 − 449 = 495. Since 146 and 449 have the same size and the same digit-difference, and yield the same difference in the reverse order procedure, then Discovery 7 holds.

EXAMPLE 11: The numbers 90007 and 28884 have the same size of 5 and the same digit-difference of 2. Applying the reverse order procedure to 90007 gives the difference as 90007 − 70009 = 19998. Applying the reverse order procedure to 28884 gives the difference as 48882 − 28884 = 19998. Since 90007 and 28884 have the same size and the same digit-difference, and yield the same difference in the reverse order procedure, then Discovery 7 holds.

CHAPTER THREE

Primary Number Patterns and Primary Sequences

1. Preliminaries

Definition 1: A primary number pattern is an infinite number pattern of the form

$$a, ac, abc, abbc, abbbc \ldots$$

where a, b, and c are digits, $a \neq 0$, $|a - c| = 1$, and $|a - c|$ is the absolute value of the difference between a and c.

If b = a, then the form of the infinite number pattern becomes

$$a, ac, aac, aaac, aaaac \ldots$$

If b = c, then the form of the infinite number pattern becomes

$$a, ac, acc, accc, acccc \ldots$$

If a = 1, and b = c = 0, then the form of the infinite number pattern becomes

$$1, 10, 100, 1000, 10000 \ldots$$

The preceding number pattern is easy to remember because its terms are associated with powers of 10 given by

$$10^{k-1}$$

for k = 1, 2 . . . , where

$$10^0 = 1.$$

The number pattern 1, 10, 100, 1000, 10000 . . . is also the preferred number pattern to use in investigating infinite sequences of multiples of 9 that are based on primary number patterns.

We refer to the number pattern 1, 10, 100, 1000, 10000 . . . as the **standard primary number pattern**. We call any other primary number pattern a **nonstandard primary number pattern.**

EXAMPLE 1: Three of the nonstandard primary number patterns that have 4 as their first term are

4, 43, 403, 4003, 40003 . . . , 4, 43, 443, 4443, 44443 . . . , and 4, 43, 433, 4333, 43333 . . .

EXAMPLE 2: Three of the nonstandard primary number patterns that have 7 as their first term are

7, 78, 798, 7998, 79998 . . . , 7, 78, 778, 7778, 77778 . . . , and 7, 78, 788, 7888, 78888 . . .

EXAMPLE 3: Three of the nonstandard primary number patterns that have 1 as their first term are

1, 10, 110, 1110, 11110 . . . , 1, 12, 112, 1112, 11112 . . . , and 1, 12, 122, 1222, 12222 . . .

Definition 2: A primary sequence is an infinite sequence of multiples of 9 that is formed if the reverse order procedure is applied to a primary number pattern and the given multiples of 9 are arranged in ascending order.

Definition 3: Applying the reverse order procedure to a given primary number pattern means that the reverse order procedure is applied to the terms of the given primary number pattern in a sequential manner.

Assumption: Applying the reverse order procedure to the terms of a given primary number pattern can be continued indefinitely.

EXAMPLE 4: Suppose the given primary number pattern is

$$4, 43, 403, 4003, 40003 \ldots$$

Then applying the reverse order procedure to this number pattern gives the differences as

$$4 - 4 = 0,\ 43 - 34 = 9,\ 403 - 304 = 99,\ 4003 - 3004 = 999 \ldots$$

Arranging the given multiples of 9 in ascending order yields a primary sequence

$$0, 9, 99, 999 \ldots$$

EXAMPLE 5: Suppose the given primary number pattern is

$$1, 10, 100, 1000, 10000 \ldots$$

Then applying the reverse order procedure to this number pattern gives the differences as

$$1 - 1 = 0,\ 10 - 01 = 9,\ 100 - 001 = 99,\ 1000 - 0001 = 999 \ldots$$

Arranging the given multiples of 9 in ascending order yields a primary sequence

$$0, 9, 99, 999 \ldots$$

EXAMPLE 6: Suppose the given primary number pattern is

$$7, 78, 788, 7888, 78888 \ldots$$

Then applying the reverse order procedure to this number pattern gives the differences as

$$7 - 7 = 0,\ 87 - 78 = 9,\ 887 - 788 = 99,\ 8887 - 7888 = 999 \ldots$$

Arranging the given multiples of 9 in ascending order yields a primary sequence

$$0, 9, 99, 999 \ldots$$

We generalize from the preceding three examples that applying the reverse order procedure to any primary number pattern and arranging the given multiples of 9 in ascending order, yields a primary sequence

$$0, 9, 99, 999 \ldots$$

whose n^{th} term is

$$10^{n-1} - 1$$

for $n = 1, 2 \ldots$

We refer to the primary sequence $0, 9, 99, 999 \ldots$ as **the primary sequence on digits of a number.**

Definition 4: A pattern modifier of a given primary number pattern is a nonzero digit that adjoins terms of the given primary number pattern in a sequential manner to form terms of a new infinite number pattern.

Definition 5: Applying a pattern modifier to a given primary number pattern means adjoining a pattern modifier and terms of the given primary number pattern in a sequential manner.

Assumption: Adjoining a pattern modifier and terms of a given primary number pattern in a sequential manner can be continued indefinitely.

The choice of a pattern modifier of a given primary number pattern is arbitrary.

The following example gives meaning to the idea of applying a pattern modifier to a primary number pattern.

EXAMPLE 7: Consider the standard primary number pattern

$$1, 10, 100, 1000, 10000 \ldots$$

Let 5 be the pattern modifier of the given primary number pattern. Then new infinite number patterns arise as follows.

Suppose 5 is applied 1 time to the given primary number pattern. Then this application produces a new infinite number pattern

$$515, 5105, 51005, 510005 \ldots$$

Suppose 5 is applied 2 times to the given primary number pattern. Then this application produces a new infinite number pattern

$$55155, 551055, 5510055, 55100055 \ldots$$

Suppose 5 is applied 3 times to the given primary number pattern. Then this application produces a new infinite number pattern

$$5551555, 55510555, 555100555, 5551000555 \ldots$$

Applying 5 to the given primary number pattern in the manner indicated can be done as many times as desired.

Discovery: The nth term of an infinite sequence of multiples of 9 which are generated with the reverse order procedure from an infinite number pattern that is formed if a pattern modifier is applied k times to a primary number pattern, is 10^k as large as the nth term of the primary sequence on digits of a number.

2. **The nth term of an infinite sequence of multiples of 9 which are generated with the reverse order procedure from a number pattern that is formed if a pattern modifier is applied k times to a primary number pattern**

The nth term of the stated infinite sequence of multiples of 9 is based on the standard primary number pattern

$$1, 10, 100, 1000, 10000 \ldots$$

Let 7 be the pattern modifier of the given primary number pattern. Then infinite sequences of multiples of 9 arise as follows.

Suppose 7 is not applied to the given primary number pattern. Then the given primary number pattern remains the same.

Applying the reverse order procedure to the given primary number pattern gives the differences as

$$1 - 1 = 0,\ 10 - 01 = 9,\ 100 - 001 = 99 \ldots$$

Arranging the given multiples of 9 in ascending order yields the primary sequence on digits of a number

$$0, 9, 99 \ldots$$

whose n^{th} term is

$$10^{n-1} - 1$$

for $n = 1, 2 \ldots$

Suppose 7 is applied 1 time to the given primary number pattern. Then this application produces a new infinite number pattern

$$717, 7107, 71007 \ldots$$

Applying the reverse order procedure to this number pattern gives the differences as

$$717 - 717 = 0,\ 7107 - 7017 = 90,\ 71007 - 70017 = 990 \ldots$$

Arranging the given multiples of 9 in ascending order yields an infinite sequence

$$0, 90, 990 \ldots$$

whose n^{th} term is

$$10(10^{n-1} - 1)$$

for $n = 1, 2 \ldots$

Suppose 7 is applied 2 times to the given primary number pattern. Then this application produces a new infinite number pattern

$$77177, 771077, 7710077 \ldots$$

Applying the reverse order procedure to this number pattern gives the differences as

$$77177 - 77177 = 0,\ 771077 - 770177 = 900,\ 7710077 - 7700177 = 9900\ \ldots$$

Arranging the given multiples of 9 in ascending order yields an infinite sequence

$$0, 900, 9900 \ldots$$

whose n^{th} term is

$$10^2(10^{n-1} - 1)$$

for $n = 1, 2 \ldots$

We generalize from the three preceding n^{th} terms that the nth term of the stated infinite sequence of multiples of 9 is

$$10^k(10^{n-1} - 1)$$

for $n = 1, 2 \ldots$, and $k = 0, 1, 2 \ldots$

It follows that $10^k(10^{n-1} - 1)$, the nth term of an infinite sequence of multiples of 9 which are generated with the reverse order procedure from a number pattern that is formed if a pattern modifier is applied k times to a primary number pattern, is 10^k as large as $10^{n-1} - 1$, the nth term of the primary sequence on digits of a number. Hence the given Discovery holds.

3. **The nth term of an infinite sequence of multiples of 9 which are generated with the reverse order procedure from a number pattern that remains if the first k terms are dropped from a primary number pattern**

The nth term of the stated infinite sequence of multiples of 9 is based on the standard primary number pattern

$$1, 10, 100, 1000, 10000 \ldots$$

Suppose no term is dropped from the given primary number pattern. Then the given primary number pattern remains the same.

Applying the reverse order procedure to the given primary number pattern gives the differences as

$$1 - 1 = 0,\ 10 - 01 = 9,\ 100 - 001 = 99 \ldots$$

Arranging the given multiples of 9 in ascending order yields the primary sequence on digits of a number

$$0, 9, 99 \ldots$$

whose n^{th} term is

$$10^{n-1} - 1$$

for $n = 1, 2 \ldots$

Suppose the first term is dropped from the given primary number pattern. Then the remaining number pattern is

$$10, 100, 1000, 10000 \ldots$$

Applying the reverse order procedure to this number pattern gives the differences as

$$10 - 01 = 9,\ 100 - 001 = 99,\ 1000 - 0001 = 999 \ldots$$

Arranging the given multiples of 9 in ascending order yields an infinite sequence

$$9, 99, 999 \ldots$$

whose n^{th} term is

$$10^n - 1$$

for $n = 1, 2 \ldots$

Suppose the first 2 terms are dropped from the given primary number pattern. Then the remaining number pattern is

$$100, 1000, 10000, 100000 \ldots$$

Applying the reverse order procedure to this number pattern give the differences as

$$100 - 001 = 99,\ 1000 - 0001 = 999,\ 10000 - 00001 = 9999 \ldots$$

Arranging the given multiples of 9 in ascending order yields an infinite sequence

$$99, 999, 999 \ldots$$

whose n^{th} term is

$$10^{n+1} - 1$$

for $n = 1, 2 \ldots$

We generalize from the three preceding n^{th} terms that the nth term of the stated infinite sequence of multiples of 9 is

$$10^{n+k-1} - 1$$

for $n = 1, 2 \ldots$, and $k = 0, 1, 2 \ldots$

CHAPTER FOUR

Subprimary Number Patterns and Subprimary Sequences

1. Preliminaries

Definition 1: A subprimary number pattern is an infinite number pattern whose terms are determined by its natural order of terms, an infinite arithmetic sequence of positive integers with specified first term and common difference whose terms indicate which terms of a given primary number pattern in their sequential order, are terms of a subprimary number pattern.

We construct a subprimary number pattern by first specifying its natural order of terms. We then determine from the natural order of terms which terms of a given primary number pattern in their sequential order, are terms of a subprimary number pattern.

EXAMPLE 1: Consider the standard primary number pattern

$$1, 10, 100, 1000, 10000 \ldots$$

Suppose the natural order of terms of its subprimary number pattern has first term 1 and common difference 2. Then the natural order of terms of the subprimary number pattern is

$$1, 3, 5 \ldots$$

We interpret this natural order of terms as follows: the terms of the given primary number pattern which are terms of a subprimary number pattern are, in their sequential order, the 1st term, the 3rd term, the 5th term, and so on.

Hence the subprimary number pattern is

$$1, 100, 10000 \ldots$$

EXAMPLE 2: Consider the standard primary number pattern

$$1, 10, 100, 1000, 10000 \ldots$$

Suppose the natural order of terms of its subprimary number pattern has first term 2 and common difference 2. Then the natural order of terms of the subprimary number pattern is

$$2, 4, 6 \ldots$$

Hence the subprimary number pattern is

$$10, 1000, 100000 \ldots$$

EXAMPLE 3: Consider a nonstandard primary number pattern

$$2, 23, 203, 2003, 20003 \ldots$$

Suppose the natural order of terms of its subprimary number pattern has first term 1 and common difference 3. Then the natural order of terms of the subprimary number pattern is

$$1, 4, 7 \ldots$$

Hence the subprimary number pattern is

$$2, 2003, 2000003 \ldots$$

Definition 2: A subprimary sequence is an infinite sequence of multiples of 9 that is formed if the reverse order procedure is applied to a subprimary number pattern and the given multiples of 9 are arranged in ascending order.

Definition 3: Applying the reverse order procedure to a given subprimary number pattern means that the reverse order procedure is applied to the terms of the given subprimary number pattern in a sequential manner.

Assumption: Applying the reverse order procedure to the terms of a given subprimary number pattern can be continued indefinitely.

EXAMPLE 4: Consider the standard primary number pattern

$$1, 10, 100, 1000, 10000 \ldots$$

Suppose the natural order of terms of its subprimary number pattern is

$$1, 4, 7 \ldots$$

Then the subprimary number pattern is

$$1, 1000, 1000000 \ldots$$

Applying the reverse order procedure to this number pattern gives the differences as

$$1 - 1 = 0,\ 1000 - 0001 = 999,\ 1000000 - 0000001 = 999999 \ldots$$

Arranging the given multiples of 9 in ascending order yields a subprimary sequence

$$0, 999, 999999 \ldots$$

EXAMPLE 5: Consider a nonstandard primary number pattern

$$7, 78, 728, 7228, 72228 \ldots$$

Suppose the natural order of terms of its subprimary number pattern is

$$1, 3, 5 \ldots$$

Then the subprimary number pattern is

$$7, 728, 72228 \ldots$$

Applying the reverse order procedure to this number pattern gives the differences as

$$7 - 7 = 0,\ 827 - 728 = 99,\ 82227 - 72228 = 9999 \ldots$$

Arranging the given multiples of 9 in ascending order yields a subprimary sequence

$$0, 99, 9999 \ldots$$

Discovery 1: Two or more subprimary number patterns that have the same natural order of terms yield the same difference if the reverse order procedure is applied to each of them.

EXAMPLE 6: Consider the standard primary number pattern

$$1, 10, 100, 1000, 10000 \ldots$$

Suppose the natural order of terms of its subprimary number pattern is

$$1, 3, 5 \ldots$$

Then the subprimary number pattern is

$$1, 100, 10000 \ldots$$

Applying the reverse order procedure to this number pattern and arranging the given multiples of 9 in ascending order, yields a subprimary sequence

$$0, 99, 9999 \ldots$$

Now consider a nonstandard primary number pattern

$$7, 78, 728, 7228, 72228 \ldots$$

Suppose the natural order of terms of its subprimary number pattern is

$$1, 3, 5 \ldots$$

Then the subprimary number pattern is

$$7, 728, 72228 \ldots$$

Applying the reverse order procedure to this number pattern and arranging the given multiples of 9 in ascending order, yields a subprimary sequence

$$0, 99, 9999 \ldots$$

Since the two given subprimary number patterns have the same natural order of terms and yield the same subprimary sequence in the reverse order procedure, then Discovery 1 holds.

Definition 4: A pattern modifier of a given subprimary number pattern is a nonzero digit that adjoins terms of the given subprimary number pattern in a sequential manner to form terms of a new infinite number pattern.

Definition 5: Applying a pattern modifier to a given subprimary number pattern means adjoining a pattern modifier and terms of the given subprimary number pattern in a sequential manner.

Assumption: Adjoining a pattern modifier and terms of a given subprimary number pattern in a sequential manner can be continued indefinitely.

The choice of a pattern modifier of a given subprimary number pattern is arbitrary.

The following example gives meaning to the idea of applying a pattern modifier to a subprimary number pattern.

EXAMPLE 7: Consider the standard primary number pattern

$$1, 10, 100, 1000, 10000 \ldots$$

Suppose the natural order of terms of its subprimary number pattern is

$$1, 4, 7 \ldots$$

Then the subprimary number pattern is

$$1, 1000, 1000000 \ldots$$

Let 2 be the pattern modifier of the given subprimary number pattern. Then new infinite number patterns arise as follows.

Suppose 2 is applied 1 time to the given subprimary number pattern. Then this application produces a new infinite number pattern

$$212, 210002, 210000002 \ldots$$

Suppose 2 is applied 2 times to the given subprimary number pattern. Then this application produces a new infinite number pattern

$$22122, 22100022, 22100000022 \ldots$$

Suppose 2 is applied 3 times to the given subprimary number pattern. Then this application produces a new infinite number pattern

$$2221222, 2221000222, 2221000000222 \ldots$$

Applying 2 to the given subprimary number pattern in the manner indicated can be done as many times as desired.

Discovery 2: The nth term of an infinite sequence of multiples of 9 which are generated with the reverse order procedure from a number pattern that is formed if a pattern modifier is applied y times to a given subprimary number pattern, is 10^y as large as the nth term of a subprimary sequence that is generated from the given subprimary number pattern.

2. **The nth term of a subprimary sequence that is generated from a subprimary number pattern whose natural order of terms has first term 1 and common difference 2k, where k is a positive integer**

The nth term of the stated subprimary sequence is based on the standard primary number pattern

$$1, 10, 100, 1000, 10000 \ldots$$

Suppose the natural order of terms of its subprimary number pattern has first term 1 and common difference 2. Then the natural order of terms of the subprimary number pattern is

$$1, 3, 5 \ldots$$

Hence the subprimary number pattern is

$$1, 100, 10000 \ldots$$

Applying the reverse order procedure to this number pattern and arranging the given multiples of 9 in ascending order, yields a subprimary sequence

$$0, 99, 9999 \ldots$$

whose n^{th} term is

$$10^{2n-2} - 1$$

for $n = 1, 2 \ldots$

For the given primary number pattern, suppose the natural order of terms of its subprimary number pattern has first term 1 and common difference 4. Then the natural order of terms of the subprimary number pattern is

$$1, 5, 9 \ldots$$

Hence the subprimary number pattern is

$$1, 10000, 100000000 \ldots$$

Applying the reverse order procedure to this number pattern and arranging the given multiples of 9 in ascending order, yields a subprimary sequence

$$0, 9999, 99999999 \ldots$$

whose n^{th} term is

$$10^{4n-4} - 1$$

for $n = 1, 2 \ldots$

We generalize from the two preceding n^{th} terms that the nth term of the stated subprimary sequence is

$$10^{2nk-2k} - 1$$

for $n = 1, 2 \ldots$, and $k = 1, 2 \ldots$

3. **The nth term of a subprimary sequence that is generated from a subprimary number pattern whose natural order of terms has first term 1 and common difference 2k + 1, where k is a positive integer**

The nth term of the stated subprimary sequence is based on the standard primary number pattern

$$1, 10, 100, 1000, 10000 \ldots$$

Suppose the natural order of terms of its subprimary number pattern has first term 1 and common difference 3. Then the natural order of terms of the subprimary number pattern is

$$1, 4, 7 \ldots$$

Hence the subprimary number pattern is

$$1, 1000, 1000000 \ldots$$

Applying the reverse order procedure to this number pattern and arranging the given multiples of 9 in ascending order, yields a subprimary sequence

$$0, 999, 999999 \ldots$$

whose n^{th} term is

$$10^{3n-3} - 1$$

for n = 1, 2 . . .

For the given primary number pattern, suppose the natural order of terms of its subprimary number pattern has first term 1 and common difference 5. Then the natural order of terms of the subprimary number pattern is

$$1, 6, 11 \ldots$$

Hence the subprimary number pattern is

$$1, 100000, 10000000000 \ldots$$

Applying the reverse order procedure to this number pattern and arranging the given multiples of 9 in ascending order, yields a subprimary sequence

$$0, 99999, 9999999999 \ldots$$

whose nth term is

$$10^{5n-5} - 1$$

for n = 1, 2 ...

We generalize from the two preceding nth terms that the nth term of the stated subprimary sequence is

$$10^{(2k+1)n - (2k+1)} - 1$$

for n = 1, 2 ..., and k = 1, 2 ...

4. **The nth term of a subprimary sequence that is generated from a subprimary number pattern whose natural order of terms has first term 2 and common difference 2k, where k is a positive integer**

The nth term of the stated subprimary sequence is based on the standard primary number pattern

$$1, 10, 100, 1000, 10000 \ldots$$

Suppose the natural order of terms of its subprimary number pattern has first term 2 and common difference 2. Then the natural order of terms of the subprimary number pattern is

$$2, 4, 6 \ldots$$

Hence the subprimary number pattern is

$$10, 1000, 100000 \ldots$$

Applying the reverse order procedure to this number pattern and arranging the given multiples of 9 in ascending order, yields a subprimary sequence

$$9, 999, 99999 \ldots$$

whose nth term is

$$10^{2n-1} - 1$$

for n = 1, 2 . . .

For the given primary number pattern, suppose the natural order of terms of its subprimary number pattern has first term 2 and common difference 4. Then the natural order of terms of the subprimary number pattern is

$$2, 6, 10 \ldots$$

Hence the subprimary number pattern is

$$10, 100000, 1000000000 \ldots$$

Applying the reverse order procedure to this number pattern and arranging the given multiples of 9 in ascending order, yields a subprimary sequence

$$9, 99999, 999999999 \ldots$$

whose nth term is

$$10^{4n-3} - 1$$

for n = 1, 2 . . .

We generalize from the two preceding nth terms that the nth term of the stated subprimary sequence is

$$10^{2kn - 2k + 1} - 1$$

for n = 1, 2 . . . , and k = 1, 2 . . .

5. **The nth term of a subprimary sequence that is formed from a subprimary number pattern whose natural order of terms has first**

term 2 and common difference 2k + 1, where k is a positive integer

The nth term of the stated subprimary sequence is based on the standard primary number pattern

$$1, 10, 100, 1000, 10000 \ldots$$

Suppose the natural order of terms of its subprimary number pattern has first term 2 and common difference 3. Then the natural order of terms of the subprimary number pattern is

$$2, 5, 8 \ldots$$

Hence the subprimary number pattern is

$$10, 10000, 10000000 \ldots$$

Applying the reverse order procedure to this number pattern and arranging the given multiples of 9 in ascending order, yields a subprimary sequence

$$9, 9999, 9999999 \ldots$$

whose n^{th} term is

$$10^{3n-2} - 1$$

for n = 1, 2 . . .

For the given primary number pattern, suppose the natural order of terms of its subprimary number pattern has first term 2 and common difference 5. Then the natural order of terms of the subprimary number pattern is

$$2, 7, 12 \ldots$$

Hence the subprimary number pattern is

$$10, 1000000, 100000000000 \ldots$$

Applying the reverse order procedure to this number pattern and arranging the given multiples of 9 in ascending order, yields a subprimary sequence

$$9, 999999, 99999999999 \ldots$$

whose n^{th} term is

$$10^{5n-4} - 1$$

for $n = 1, 2 \ldots$, and $k = 1, 2 \ldots$

We generalize from the two preceding n^{th} terms that the nth term of the stated subprimary sequence is

$$10^{(2k+1)n - 2k} - 1$$

for $n = 1, 2 \ldots$, and $k = 1, 2 \ldots$

6. **The nth term of a subprimary sequence that is generated from a subprimary number pattern whose natural order of terms has first term 3 and common difference 3k, where k is a positive integer**

The nth term of the stated subprimary sequence is based on the standard primary number pattern

$$1, 10, 100, 1000, 10000 \ldots$$

Suppose the natural order of terms of its subprimary number pattern has first term 3 and common difference 3. Then the natural order of terms of the subprimary number pattern is

$$3, 6, 9 \ldots$$

Hence the subprimary number pattern is

$$100, 100000, 100000000 \ldots$$

Applying the reverse order procedure to this number pattern and arranging the given multiples of 9 in ascending order, yields a subprimary sequence

$$99, 99999, 99999999 \ldots$$

whose n^{th} term is

$$10^{3n-1} - 1$$

for $n = 1, 2 \ldots$

For the given primary number pattern, suppose the natural order of terms of its subprimary number pattern has first term 3 and common difference 6. Then the natural order of terms of the subprimary number pattern is

$$3, 9, 15 \ldots$$

Hence the subprimary number pattern is

$$100, 100000000, 100000000000000 \ldots$$

Applying the reverse order procedure to this number pattern and arranging the given multiples of 9 in ascending order, yields a subprimary sequence

$$99, 99999999, 99999999999999 \ldots$$

whose n^{th} term is

$$10^{6n-4} - 1$$

for $n = 1, 2 \ldots$

We generalize from the two preceding n^{th} terms that the nth term of the stated subprimary sequence is

$$10^{3nk - 3k + 2} - 1$$

for n = 1, 2..., and k = 1, 2...

7. **The nth term of an infinite sequence of multiples of 9 which are generated with the reverse order procedure from a number pattern that is formed if a pattern modifier is applied y times to a subprimary number pattern whose natural order of terms has first term 1 and common difference 2k, where k is a positive integer**

The nth term of the stated infinite sequence of multiples of 9 is based on the standard primary number pattern

$$1, 10, 100, 1000, 10000 \ldots$$

Suppose the natural order of terms of its subprimary number pattern has first term 1 and common difference 2. Then the natural order of terms of the subprimary number pattern is

$$1, 3, 5 \ldots$$

Hence the subprimary number pattern is

$$1, 100, 10000 \ldots$$

Let 3 be the pattern modifier of this subprimary number pattern. Then infinite sequences of multiples of 9 arise as follows.

Suppose 3 is not applied to the given subprimary number pattern. Then the given subprimary number pattern remains the same.

Applying the reverse order procedure to this number pattern and arranging the given multiples of 9 in ascending order, yields a subprimary sequence

$$0, 99, 9999 \ldots$$

whose n^{th} term is

$$10^{2n-2} - 1$$

for $n = 1, 2 \ldots$

Suppose 3 is applied 1 time to the given subprimary number pattern. Then this application produces a new infinite number pattern

$$313, 31003, 3100003 \ldots$$

Applying the reverse order procedure to this number pattern and arranging the given multiples of 9 in ascending order, yields an infinite sequence

$$0, 990, 99990 \ldots$$

whose n^{th} term is

$$10(10^{2n-2} - 1)$$

for $n = 1, 2 \ldots$

Suppose 3 is applied 2 times to the given subprimary number pattern. Then this application produces a new infinite number pattern

$$33133, 3310033, 331000033 \ldots$$

Applying the reverse order procedure to this number pattern and arranging the given multiples of 9 in ascending order, yields an infinite sequence

$$0, 9900, 999900 \ldots$$

whose n^{th} term is

$$10^2(10^{2n-2} - 1)$$

for n = 1, 2 . . .

For the given primary number pattern, suppose the natural order of terms of its subprimary number pattern has first term 1 and common difference 4. Then the natural order of terms of the subprimary number pattern is

$$1, 5, 9 \ldots$$

Hence the subprimary number pattern is

$$1, 10000, 100000000 \ldots$$

Let 3 be the pattern modifier of this subprimary number pattern. Then infinite sequences of multiples of 9 arise as follows.

Suppose 3 is not applied to the given subprimary number pattern. Then the given subprimary number pattern remains the same.

Applying the reverse order procedure to this number pattern and arranging the given multiples of 9 in ascending order, yields a subprimary sequence

$$0, 9999, 99999999 \ldots$$

whose n^{th} term is

$$10^{4n-4} - 1$$

for n = 1, 2 . . .

Suppose 3 is applied 1 time to the given subprimary number pattern. Then this application produces a new infinite number pattern

$$313, 3100003, 31000000003 \ldots$$

Applying the reverse order procedure to this number pattern and arranging the given multiples of 9 in ascending order, yields an infinite sequence

$$0, 99990, 999999990 \ldots$$

whose n^{th} term is

$$10(10^{4n-4} - 1)$$

for $n = 1, 2 \ldots$

Suppose 3 is applied 2 times to the given subprimary number pattern. Then this application produces a new infinite number pattern

$$33133, 331000033, 3310000000033 \ldots$$

Applying the reverse order procedure to this number pattern and arranging the given multiples of 9 in ascending order, yields an infinite sequence

$$0, 999900, 9999999900 \ldots$$

whose n^{th} term is

$$10^2(10^{4n-4} - 1)$$

for $n = 1, 2 \ldots$

We generalize from the six preceding n^{th} terms that the nth term of the stated infinite sequence of multiples of 9 is

$$10^y(10^{2kn-2k} - 1)$$

for $n = 1, 2 \ldots$, $k = 1, 2 \ldots$, and $y = 0, 1, 2 \ldots$

It follows from the preceding n^{th} term that the nth term of an infinite sequence of multiples of 9 which are generated with the reverse order procedure from a number pattern that is formed if a pattern modifier is applied y times to a subprimary number pattern whose natural order of terms has first term 1 and common difference 2k, is 10^y as large as the nth term of a subprimary sequence that is generated from a subprimary number pattern whose natural order of terms has first term 1 and common difference 2k, where k is a positive integer. Hence Discovery 2 of this Chapter holds.

8. **The nth term of an infinite sequence of multiples of 9 which are generated with the reverse order procedure from a number pattern that is formed if a pattern modifier is applied y times to a subprimary number pattern whose natural order of terms has first term 2 and common difference 2k, where k is a positive integer**

The nth term of the stated infinite sequence of multiples of 9 is based on the standard primary number pattern

$$1, 10, 100, 1000, 10000 \ldots$$

Suppose the natural order of terms of its subprimary number pattern has first term 2 and common difference 2. Then the natural order of terms of the subprimary number pattern is

$$2, 4, 6 \ldots$$

Hence the subprimary number pattern is

$$10, 1000, 100000 \ldots$$

Let 5 be the pattern modifier of this subprimary number pattern. Then infinite sequences of multiples of 9 arise as follows.

Suppose 5 is not applied to the given subprimary number pattern. Then the given subprimary number pattern remains the same.

Applying the reverse order procedure to this number pattern and arranging the given multiples of 9 in ascending order, yields a subprimary sequence

$$9, 999, 99999 \ldots$$

whose n^th term is

$$10^{2n-1} - 1$$

for n = 1, 2 . . .

Suppose 5 is applied 1 time to the given subprimary number pattern. Then this application produces a new infinite number pattern

$$5105, 510005, 51000005 \ldots$$

Applying the reverse order procedure to this number pattern and arranging the given multiples of 9 in ascending order, yields an infinite sequence

$$90, 9990, 999990 \ldots$$

whose n^th term is

$$10(10^{2n-1} - 1)$$

for n = 1, 2 . . .

Suppose 5 is applied 2 times to the given subprimary number pattern. Then this application produces a new infinite number pattern

$$551055, 55100055, 5510000055 \ldots$$

Applying the reverse order procedure to this number pattern and arranging the given multiples of 9 in ascending order, yields an infinite sequence

$$900, 99900, 9999900 \ldots$$

whose n^th term is

$$10^2(10^{2n-1} - 1)$$

for n = 1, 2 . . .

For the given primary number pattern, suppose the natural order of terms of its subprimary number pattern has first term 2 and common difference 4. Then the natural order of terms of the subprimary number pattern is

$$2, 6, 10 \ldots$$

Hence the subprimary number pattern is

$$10, 100000, 1000000000 \ldots$$

Let 5 be the pattern modifier of this subprimary number pattern. Then infinite sequences of multiples of 9 arise as follows.

Suppose 5 is not applied to the given subprimary number pattern. Then the given subprimary number pattern remains the same.

Applying the reverse order procedure to this number pattern and arranging the given multiples of 9 in ascending order, yields a subprimary sequence

$$9, 99999, 999999999 \ldots$$

whose n^{th} term is

$$10^{4n-3} - 1$$

for n = 1, 2 . . .

Suppose 5 is applied 1 time to the given subprimary number pattern. Then this application produces a new infinite number pattern

$$5105, 51000005, 510000000005 \ldots$$

Applying the reverse order procedure to this number pattern and arranging the given multiples of 9 in ascending order, yields an infinite sequence

$$90, 999990, 9999999990 \ldots$$

whose n^{th} term is

$$10(10^{4n-3} - 1)$$

for $n = 1, 2 \ldots$

Suppose 5 is applied 2 times to the given subprimary number pattern. Then this application produces a new infinite number pattern

$$551055, 5510000055, 55100000000055 \ldots$$

Applying the reverse order procedure to this number pattern and arranging the given multiples of 9 in ascending order, yields an infinite sequence

$$900, 9999900, 99999999900 \ldots$$

whose n^{th} term is

$$10^2(10^{4n-3} - 1)$$

for $n = 1, 2 \ldots$

We generalize from the six preceding n^{th} terms that the nth term of the stated infinite sequence of multiples of 9 is

$$10^y(10^{2kn - 2k + 1} - 1)$$

for $n = 1, 2 \ldots, k = 1, 2 \ldots,$ and $y = 0, 1, 2 \ldots$

It follows from the preceding n^{th} term that the nth term of an infinite sequence of multiples of 9 which are generated with the reverse order procedure from a number pattern that is formed if a pattern modifier is applied y times to a subprimary number pattern whose natural order of terms has first term 2 and common difference 2k, is 10^y as large as the nth term of a subprimary sequence that is generated from a subprimary number pattern whose natural order of terms has first term 2 and common difference 2k, where k is a positive integer. Hence Discovery 2 of this Chapter holds.

9. **The nth term of an infinite sequence of multiples of 9 which are generated with the reverse order procedure from a number pattern that remains if the first y terms are dropped from a subprimary number pattern whose natural order of terms has first term 1 and common difference 2k, where k is a positive integer**

The nth term of the stated infinite sequence of multiples of 9 is based on the standard primary number pattern

$$1, 10, 100, 1000, 10000 \ldots$$

Suppose the natural order of terms of its subprimary number pattern has first term 1 and common difference 2. Then the natural order of terms of the subprimary number pattern is

$$1, 3, 5 \ldots$$

Hence the subprimary number pattern is

$$1, 100, 10000 \ldots$$

Suppose no term is dropped from the given subprimary number pattern. Then the given subprimary number pattern remains the same.

Applying the reverse order procedure to this number pattern and arranging the given multiples of 9 in ascending order, yields a subprimary sequence

$$0, 99, 9999 \ldots$$

whose n^{th} term is

$$10^{2n-2} - 1$$

for n = 1, 2 . . .

Suppose the first term is dropped from the given subprimary number pattern. Then the remaining number pattern is

$$100, 10000, 1000000 \ldots$$

Applying the reverse order procedure to this number pattern and arranging the given multiples of 9 in ascending order, yields an infinite sequence

$$99, 9999, 999999 \ldots$$

whose n^{th} term is

$$10^{2n} - 1$$

for n = 1, 2 . . .

Suppose the first 2 terms are dropped from the given subprimary number pattern. Then the remaining number pattern is

$$10000, 1000000, 100000000 \ldots$$

Applying the reverse order procedure to this number pattern and arranging the given multiples of 9 in ascending order, yields an infinite sequence

$$9999, 999999, 99999999 \ldots$$

whose n^{th} term is

$$10^{2n+2} - 1$$

for n = 1, 2 . . .

For the given primary number pattern, suppose the natural order of terms of its subprimary number pattern has first term 1 and common difference 4. Then the natural order of terms of the subprimary number pattern is

$$1, 5, 9 \ldots$$

Hence the subprimary number pattern is

$$1, 10000, 100000000 \ldots$$

Suppose no term is dropped from the given subprimary number pattern. Then the given subprimary number pattern remains the same.

Applying the reverse order procedure to this number pattern and arranging the given multiples of 9 in ascending order, yields a subprimary sequence

$$0, 9999, 99999999 \ldots$$

whose n^{th} term is

$$10^{4n-4} - 1$$

for $n = 1, 2 \ldots$

Suppose the first term is dropped from the given subprimary number pattern. Then the remaining number pattern is

$$10000, 100000000, 1000000000000 \ldots$$

Applying the reverse order procedure to this number pattern and arranging the given multiples of 9 in ascending order, yields an infinite sequence

$$9999, 99999999, 999999999999 \ldots$$

whose n^{th} term is

$$10^{4n} - 1$$

for n = 1, 2 . . .

We generalize from the five preceding n^{th} terms that the nth term of the stated infinite sequence of multiples of 9 is

$$10^{2kn - 2k + 2yk} - 1$$

for n = 1, 2 . . . , k = 1, 2 . . . , and y = 0, 1, 2 . . .

10. **The nth term of an infinite sequence of multiples of 9 which are generated with the reverse order procedure from a number pattern that remains if the first y terms are dropped from a subprimary number pattern whose natural order of terms has first term 2 and common difference 2k, where k is a positive integer**

The nth term of the stated infinite sequence of multiples of 9 is based on the standard primary number pattern

$$1, 10, 100, 1000, 10000 \ldots$$

Suppose the natural order of terms of its subprimary number pattern has first term 2 and common difference 2. Then the natural order of terms of the subprimary number pattern is

$$2, 4, 6 \ldots$$

Hence the subprimary number pattern is

$$10, 1000, 100000 \ldots$$

Suppose no term is dropped from the given subprimary number pattern. Then the given subprimary number pattern remains the same.

Applying the reverse order procedure to this number pattern and arranging the given multiples of 9 in ascending order, yields a subprimary sequence

$$9, 999, 99999 \ldots$$

whose n^{th} term is

$$10^{2n-1} - 1$$

for $n = 1, 2 \ldots$

Suppose the first term is dropped from the given subprimary number pattern. Then the remaining number pattern

$$1000, 100000, 10000000 \ldots$$

Applying the reverse order procedure to this number pattern and arranging the given multiples of 9 in ascending order, yields an infinite sequence

$$999, 99999, 9999999 \ldots$$

whose n^{th} term is

$$10^{2n+1} - 1$$

for $n = 1, 2 \ldots$

Suppose the first 2 terms are dropped from the given subprimary number pattern. Then the remaining number pattern is

$$100000, 10000000, 1000000000 \ldots$$

Applying the reverse order procedure to this number pattern and arranging the given multiples of 9 in ascending order, yields an infinite sequence

$$99999, 9999999, 999999999 \ldots$$

whose nth term is

$$10^{2n+3} - 1$$

for n = 1, 2 . . .

For the given primary number pattern, suppose the natural order of terms of its subprimary number pattern has first term 2 and common difference 4. Then the natural order of terms of the subprimary number pattern is

$$2, 6, 10 \ldots$$

Hence the subprimary number pattern is

$$10, 100000, 1000000000 \ldots$$

Suppose no term is dropped from the given subprimary number pattern. Then the given subprimary number pattern remains the same.

Applying the reverse order procedure to this number pattern and arranging the given multiples of 9 in ascending order, yields a subprimary sequence

$$9, 99999, 999999999 \ldots$$

whose nth term is

$$10^{4n-3} - 1$$

for n = 1, 2 . . .

Suppose the first term is dropped from the given subprimary number pattern. Then the remaining number pattern is

$$100000, 1000000000, 10000000000000 \ldots$$

Applying the reverse order procedure to this number pattern and arranging the given multiples of 9 in ascending order, yields an infinite sequence

$$99999, 999999999, 9999999999999 \ldots$$

whose n^{th} term is

$$10^{4n+1} - 1$$

for n = 1, 2 ...

We generalize from the five preceding n^{th} terms that the nth term of the stated infinite sequence of multiples of 9 is

$$10^{2kn - 2k + 1 + 2yk} - 1$$

for n = 1, 2 ..., k = 1, 2 ..., and y = 0, 1, 2 ...

11. **The nth term of an infinite sequence of multiples of 9 which are generated with the reverse order procedure from a number pattern that remains if the first y terms are dropped from a subprimary number pattern whose natural order of terms has first term 2 and common difference 2k + 1, where k is a positive integer**

The nth term of the stated infinite sequence of multiples of 9 is based on the standard primary number pattern

$$1, 10, 100, 1000, 10000 \ldots$$

Suppose the natural order of terms of its subprimary number pattern has first term 2 and common difference 3. Then the natural order of terms of the subprimary number pattern is

$$2, 5, 8 \ldots$$

Hence the subprimary number pattern is

$$10, 10000, 10000000 \ldots$$

Suppose no term is dropped from the given subprimary number pattern. Then the given subprimary number pattern remains the same.

Applying the reverse order procedure to this number pattern and arranging the given multiples of 9 in ascending order, yields a subprimary sequence

$$9, 9999, 9999999 \ldots$$

whose n^{th} term is

$$10^{3n-2} - 1$$

for n = 1, 2 . . .

Suppose the first term is dropped from the given subprimary number pattern. Then the remaining number pattern

$$10000, 10000000, 10000000000 \ldots$$

Applying the reverse order procedure to this number pattern and arranging the given multiples of 9 in ascending order, yields an infinite sequence

$$9999, 9999999, 9999999999 \ldots$$

whose n^{th} term is

$$10^{3n+1} - 1$$

for n = 1, 2 . . .

Suppose the first 2 terms are dropped from the given subprimary number pattern. Then the remaining number pattern is

$$10000000, 10000000000, 10000000000000 \ldots$$

Applying the reverse order procedure to this number pattern and arranging the given multiples of 9 in ascending order, yields an infinite sequence

$$9999999, 9999999999, 9999999999999 \ldots$$

whose n^{th} term is

$$10^{3n+4} - 1$$

for $n = 1, 2 \ldots$

For the given primary number pattern, suppose the natural order of terms of its subprimary number pattern has first term 2 and common difference 5. Then the natural order of terms of the subprimary number pattern is

$$2, 7, 12 \ldots$$

Hence the subprimary number pattern is

$$10, 1000000, 100000000000 \ldots$$

Suppose no term is dropped from the given subprimary number pattern. Then the given subprimary number pattern remains the same.

Applying the reverse order procedure to this number pattern and arranging the given multiples of 9 in ascending order, yields a subprimary sequence

$$9, 999999, 99999999999 \ldots$$

whose n^{th} term is

$$10^{5n-4} - 1$$

for $n = 1, 2 \ldots$

Suppose the first term is dropped from the given subprimary number pattern. Then the remaining number pattern is

$$1000000, 100000000000, 10000000000000000 \ldots$$

Applying the reverse order procedure to this number pattern and arranging the given multiples of 9 in ascending order, yields an infinite sequence

$$999999, 99999999999, 9999999999999999 \ldots$$

whose n^{th} term is

$$10^{5n+1} - 1$$

for $n = 1, 2 \ldots$

We generalize from the five preceding n^{th} terms that the nth term of the stated infinite sequence of multiples of 9 is

$$10^{(2k+1)n - 2k + (2k+1)y} - 1$$

for $n = 1, 2 \ldots, k = 1, 2 \ldots,$ and $y = 0, 1, 2 \ldots$

CHAPTER FIVE

Secondary Number Patterns and Secondary Sequences

1. Preliminaries

Definition 1: A secondary number pattern is an infinite number pattern of the form

$$ab, cabc, ccabcc, cccabccc \ldots$$

where a, b, and c are digits, $a \neq 0$, $c \neq 0$, $|a - b| = 1$, and $|a - b|$ is the absolute value of the difference between a and b.

If b = c, then the form of the infinite number pattern becomes

$$ac, cacc, ccaccc, cccacccc \ldots$$

If a = c, then the form of the infinite number pattern becomes

$$cb, ccbc, cccbcc, ccccbccc \ldots$$

If a = 1, b = 0, and c = 1, then the form of the infinite number pattern becomes

$$10, 1101, 111011, 11110111 \ldots$$

The preceding secondary number pattern is the only secondary number pattern which is constructed entirely from 0 and 1, the two smallest digits. The unique construction makes the preceding secondary number pattern easy to remember. This secondary number pattern is also the preferred number pattern to use in investigating infinite sequences of multiples of 9 which are based on secondary number patterns.

We refer to the secondary number pattern 10, 1101, 111011, 11110111 . . . as the **standard secondary number pattern**. We call any other secondary number pattern a **nonstandard secondary number pattern**.

EXAMPLE 1: Three of the nonstandard secondary number patterns that have 12 as their first term are

12, 7127, 771277, 77712777 . . . , 12, 2122, 221222, 22212222 . . . , and 12, 1121, 111211, 11112111 . . .

EXAMPLE 2: Three of the nonstandard secondary number patterns that have 54 as their first term are

54, 8548, 885488, 88854888 . . . , 54, 4544, 445444, 44454444 . . . , and 54, 5545, 555455, 55554555 . . .

EXAMPLE 3: Three of the nonstandard secondary number patterns that have 10 as their first term are

10, 2102, 221022, 22210222 . . . , 10, 9109, 991099, 99910999 . . . , and 10, 6106, 661066, 66610666 . . .

Definition 2: A secondary sequence is an infinite sequence of multiples of 9 that is formed if the reverse order procedure is applied to a secondary number pattern and the given multiples of 9 are arranged in ascending order.

Definition 3: Applying the reverse order procedure to a given secondary number pattern means that the reverse order procedure is applied to the terms of the given secondary number pattern in a sequential manner.

Assumption: Applying the reverse order procedure to the terms of a given secondary number pattern in a sequential manner can be continued indefinitely.

EXAMPLE 4: Suppose the given secondary number pattern is

$$89, 4894, 448944, 44489444 \ldots$$

Then applying the reverse order procedure to this number pattern gives the differences as

$$98 - 89 = 9, 4984 - 4894 = 90, 449844 - 448944 = 900 \ldots$$

Arranging the given multiples of 9 in ascending order yields a secondary sequence

$$9, 90, 900 \ldots$$

EXAMPLE 5: Suppose the given secondary number pattern is

$$32, 6326, 663266, 66632666 \ldots$$

Then applying the reverse order procedure to this number pattern gives the differences as

$$32 - 23 = 9, 6326 - 6236 = 90, 663266 - 662366 = 900 \ldots$$

Arranging the given multiples of 9 in ascending order yields a secondary sequence

$$9, 90, 900 \ldots$$

EXAMPLE 6: Suppose the given secondary number pattern is

$$10, 1101, 111011, 11110111 \ldots$$

Then applying the reverse order procedure to this number pattern gives the differences as

$$10 - 01 = 9,\ 1101 - 1011 = 90,\ 111011 - 110111 = 900\ldots$$

Arranging the given multiples of 9 in ascending order yields a secondary sequence

$$9,\ 90,\ 900\ldots$$

We generalize from the three preceding examples that applying the reverse order procedure to any secondary number pattern and arranging the given multiples of 9 in ascending order, yields a secondary sequence

$$9,\ 90,\ 900\ldots$$

whose n^{th} term is

$$9(10^{n-1})$$

for $n = 1, 2 \ldots$

We refer to the secondary sequence $9, 90, 900 \ldots$ as **the secondary sequence on digits of a number.**

Definition 4: A pattern modifier of a given secondary number pattern is a nonzero digit that adjoins the terms of the given secondary number pattern to form a new infinite number pattern.

Definition 5: Applying a pattern modifier to a given secondary number pattern means adjoining a pattern modifier and terms of the given secondary number pattern in a sequential manner.

Assumption: Adjoining a pattern modifier and terms of a given secondary number pattern in a sequential manner can be continued indefinitely.

The choice of a pattern modifier of a given secondary number pattern is arbitrary.

The following example gives meaning to the idea of applying a pattern modifier to a given secondary number pattern.

EXAMPLE 7: Consider the standard secondary number pattern

$$10, 1101, 111011, 11110111\ldots$$

Let 4 be the pattern modifier of the given secondary number pattern. Then new infinite number patterns arise as follows.

Suppose 4 is applied 1 time to the given secondary number pattern. Then this application produces a new infinite number pattern

$$4104, 411014, 41110114, 4111101114\ldots$$

Suppose 4 is applied 2 times to the given secondary number pattern. Then this application produces a new infinite number pattern

$$441044, 44110144, 4411101144, 441111011144\ldots$$

Suppose 4 is applied 3 times to the given secondary number pattern. Then this application produces a new infinite number pattern

$$44410444, 4441101444, 444111011444, 44411110111444\ldots$$

Applying 4 to the given secondary number pattern in the manner indicated can be done as many times as desired.

Discovery 1: The nth term of an infinite sequence of multiples of 9 which are generated with the reverse procedure from a number pattern that is formed if a pattern modifier is applied k times to a secondary number pattern, is 10^k as large as the nth term of the secondary sequence on digits of a number.

Discovery 2: The nth term of an infinite sequence of multiples of 9 which are generated with the reverse order procedure from a number pattern that is formed if a pattern modifier is applied k times to a secondary number pattern, is the same as the nth term of an infinite sequence of multiples of 9 which are generated with the reverse order procedure from a number pattern that remains if the first k terms are dropped from a secondary number pattern.

2. **The nth term of an infinite sequence of multiples of 9 which are generated with the reverse order procedure from a number pattern that is formed if a pattern modifier is applied k times to a secondary number pattern**

The nth term of the stated infinite sequence of multiples of 9 is based on the standard secondary number pattern

$$10, 1101, 111011, 11110111 \ldots$$

Let 6 be the pattern modifier of the given secondary number pattern. Then infinite sequences of multiples of multiples of 9 arise as follows.

Suppose 6 is not applied to the given secondary number pattern. Then the given secondary number pattern remains the same.

Applying the reverse order procedure to the given secondary number pattern gives the differences as

$$10 - 01 = 9, 1101 - 1011 = 90, 111011 - 110111 = 900 \ldots$$

Arranging the given multiples of 9 in ascending order yields the secondary sequence on digits of a number

$$9, 90, 900 \ldots$$

whose n^{th} term is

$$9(10^{n-1})$$

for n = 1, 2 ...

Suppose 6 is applied 1 time to the given secondary number pattern. Then this application produces a new infinite number pattern

$$6106, 611016, 61110116, 6111101116 \ldots$$

Applying the reverse order procedure to this number pattern gives the differences as

$$6106 - 6016 = 90, 611016 - 610116 = 900, 61110116 - 61101116 = 9000 \ldots$$

Arranging the given multiples of 9 in ascending order yields an infinite sequence

$$90, 900, 9000 \ldots$$

whose n^{th} term is

$$9(10^n)$$

for n = 1, 2 ...

Suppose 6 is applied 2 times to the given secondary number pattern. Then this application produces a new infinite number pattern

$$661066, 66110166, 6611101166, 661111011166 \ldots$$

Applying the reverse order procedure to this number pattern gives the differences as

$$661066 - 660166 = 900, 66110166 - 66101166 = 9000, 6611101166 - 6611011166 = 90000 \ldots$$

Arranging the given multiples of 9 in ascending order yields an infinite sequence

$$900, 9000, 90000 \ldots$$

whose n^{th} term is

$$9(10^{n+1})$$

for $n = 1, 2 \ldots$

We generalize from the three preceding n^{th} terms that the nth term of the stated infinite sequence of multiples of 9 is

$$9(10^{n+k-1})$$

which can be expressed as

$$10^k(9(10^{n-1}))$$

for $n = 1, 2 \ldots$, and $k = 0, 1, 2 \ldots$

It follows from the preceding n^{th} term that the nth term of an infinite sequence of multiples of 9 which are generated with the reverse order procedure from a number pattern that is formed if a pattern modifier is applied k times to a secondary number pattern, is 10^k as large as the nth term of the secondary sequence on digits of a number. Hence Discovery 1 of this Chapter holds.

3. **The nth term of an infinite sequence of multiples of 9 which are generated with the reverse order procedure from a number pattern that remains if the first k terms are dropped from a secondary number pattern**

The nth term of the stated infinite sequence of multiples of 9 is based on the standard secondary number pattern

$$10, 1101, 111011, 11110111\ldots$$

Suppose no term is dropped from the given secondary number pattern. Then the given secondary number pattern remains the same.

Applying the reverse order procedure to the to the given secondary number pattern gives the differences as

$$10 - 01 = 9,\ 1101 - 1011 = 90,\ 111011 - 110111 = 9000\ldots$$

Arranging the given multiples of 9 in ascending order yields the secondary sequence on digits of a number

$$9, 90, 9000\ldots$$

whose n^{th} term is

$$9(10^{n-1})$$

for $n = 1, 2\ldots$

Suppose the first term is dropped from the given secondary number pattern. Then the remaining number pattern is

$$1101, 111011, 11110111\ldots$$

Applying the reverse order procedure to this number pattern gives the differences as

$$1101 - 1011 = 90,\ 111011 - 110111 = 900,\ 11110111 - 11101111 = 9000$$
$$\ldots$$

Arranging the given multiples of 9 in ascending order yields an infinite sequence

$$90, 900, 9000\ldots$$

whose n^{th} term is

$$9(10^n)$$

for $n = 1, 2 \ldots$

Suppose the first 2 terms are dropped from the given secondary number pattern. Then the remaining number pattern is

$$111011, 11110111, 1111101111 \ldots$$

Applying the reverse order procedure to this number pattern give the differences as

$111011 - 110111 = 900$, $11110111 - 11101111 = 9000$, $1111101111 - 1111011111 = 90000 \ldots$

Arranging the given multiples of 9 in ascending order yields an infinite sequence

$$900, 9000, 90000 \ldots$$

whose n^{th} term is

$$9(10^{n+1})$$

for $n = 1, 2 \ldots$

We generalize from the three preceding n^{th} terms that the nth term of the stated infinite sequence of multiples of 9 is

$$9(10^{n+k-1})$$

which can be expressed as

$$10^k\{9(10^{n-1})\}$$

65

for n = 1, 2 . . . , and k = 0, 1, 2 . . .

We note that the nth term of the stated infinite sequence of multiples of 9 derived in the previous Section, is the same as the nth term of the stated infinite sequence of multiples of 9 derived in this Section.

Thus the nth term of an infinite sequence of multiples of 9 which are generated with the reverse order procedure from a number pattern that is formed if a pattern modifier is applied k times to a secondary number pattern, is the same as the nth term of an infinite sequence of multiples of 9 which are generated with the reverse order procedure from a number pattern that remains if the first k terms are dropped from a secondary number pattern. Hence Discovery 2 of this Chapter holds.

CHAPTER SIX

Subsecondary Number Patterns and Subsecondary Sequences

1. Preliminaries

Definition 1: A subsecondary number pattern is an infinite number pattern whose terms are determined by its natural order of terms, an infinite arithmetic sequence of positive integers whose terms indicate which terms of a given secondary number pattern in their sequential order, are terms of a subsecondary number pattern.

We construct a subsecondary number pattern the same way we construct a subprimary number pattern. We first specify the natural order of terms of a subsecondary number pattern. We then determine from the natural order of terms which terms of a given secondary number pattern in their sequential order, are terms of a subsecondary number pattern.

EXAMPLE 1: Consider the standard secondary number pattern

$$10, 1101, 111011, 11110111 \ldots$$

Suppose the natural order of terms of its subsecondary number pattern has first term 1 and common difference 2. Then the natural order of terms of the subsecondary number pattern is

$$1, 3, 5 \ldots$$

Hence the subsecondary number pattern is

$$10, 111011, 1111101111 \ldots$$

EXAMPLE 2: Consider the standard secondary number pattern

$$10, 1101, 111011, 11110111 \ldots$$

Suppose the natural order of terms of its subsecondary number pattern has first term 2 and common difference 2. Then the natural order of terms of the subsecondary number pattern is

$$2, 4, 6 \ldots$$

Hence the subsecondary number pattern is

$$1101, 11110111, 111111011111 \ldots$$

EXAMPLE 3: Consider a nonstandard secondary number pattern

$$12, 4124, 441244, 44412444 \ldots$$

Suppose the natural order of terms of its subsecondary number pattern has first term 1 and common difference 3. Then the natural order of terms of the subsecondary number pattern is

$$1, 4, 7 \ldots$$

Hence the subsecondary number pattern is

$$12, 44412444, 4444441244444 \ldots$$

Definition 2: A subsecondary sequence is an infinite sequence of multiples of 9 that is formed if the reverse order procedure is applied to a subsecondary number pattern and the given multiples of 9 are arranged in ascending order.

Definition 3: Applying the reverse order procedure to a given subsecondary number pattern means that the reverse order procedure is applied to the terms of the given subsecondary number pattern in a sequential manner.

Assumption: Applying the reverse order procedure to the terms of a given subsecondary number pattern can be continued indefinitely.

EXAMPLE 4: Consider the standard secondary number pattern

$$10, 1101, 111011, 11110111 \ldots$$

Suppose the natural order of terms of its subsecondary number pattern is

$$1, 4, 7 \ldots$$

Then the subsecondary number pattern is

$$10, 11110111, 11111110111111 \ldots$$

Applying the reverse order procedure to this number pattern gives the differences as

$$10 - 01 = 9, 11110111 - 11101111 = 9000, 11111110111111 = 9000000\,.$$
$$\cdot\,\cdot$$

Arranging the given multiples of 9 in ascending order yields a subsecondary sequence

$$9, 9000, 9000000 \ldots$$

EXAMPLE 5: Consider a nonstandard secondary number pattern

$$34, 1341, 113411, 11134111 \ldots$$

Suppose the natural order of terms of its subsecondary number pattern is

$$1, 3, 5 \ldots$$

Then the subsecondary number pattern is

$$34, 113411, 1111341111 \ldots$$

Applying the reverse order procedure to this number pattern gives the differences as

$$43 - 34 = 9,\ 114311 - 113411 = 900,\ 1111431111 - 1111341111 = 90000$$
$$\ldots$$

Arranging the given multiples of 9 in ascending order yields a subsecondary sequence

$$9, 900, 90000 \ldots$$

EXAMPLE 6: Consider the standard secondary number pattern

$$10, 1101, 111011, 11110111 \ldots$$

Suppose the natural order of terms of its subsecondary number pattern is

$$2, 4, 6 \ldots$$

Then the subsecondary number pattern is

$$1101, 11110111, 111111011111 \ldots$$

Applying the reverse order procedure to this number pattern gives the differences as

$$1101 - 1011 = 90,\ 11110111 - 11101111 = 9000,\ 111111011111 - 111110111111 = 900000 \ldots$$

Arranging the given multiples of 9 in ascending order yields a subsecondary sequence

$$90, 9000, 900000 \ldots$$

EXAMPLE 7: Consider the standard secondary number pattern

$$10, 1101, 111011, 11110111 \ldots$$

Suppose the natural order of terms of its subsecondary number pattern is

$$2, 5, 8 \ldots$$

Then the subsecondary number pattern is

$$1101, 1111101111, 1111111101111111 \ldots$$

Applying the reverse order procedure to this number pattern gives the differences as

$$1101 - 1011 = 90, \; 1111101111 - 1111011111 = 90000,$$
$$1111111101111111 - 1111111011111111 = 90000000 \ldots$$

Arranging the given multiples of 9 in ascending order yields a subsecondary sequence

$$90, 90000, 90000000 \ldots$$

EXAMPLE 8: Consider the standard secondary number pattern

$$10, 1101, 111011, 11110111 \ldots$$

Suppose the natural order of terms of its subsecondary number pattern is

$$3, 6, 9 \ldots$$

Then the subsecondary number pattern is

$$111011, 111111011111, 111111111011111111\ldots$$

Applying the reverse order procedure to this number pattern gives the differences as

$$111011 - 110111 = 900, 111111011111 - 111110111111 = 900000,$$
$$111111111011111111 - 111111110111111111 = 900000000\ldots$$

Arranging the given multiples of 9 in ascending order yields a subsecondary sequence

$$900, 900000, 900000000\ldots$$

Discovery 1: Two or more subsecondary number patterns that have the same natural order of terms yield the same difference if the reverse order procedure is applied to each of them.

EXAMPLE 9: Consider the standard secondary number pattern

$$10, 1101, 111011, 11110111\ldots$$

Suppose the natural order of terms of its subsecondary number pattern is

$$1, 3, 5\ldots$$

Then the subsecondary number pattern is

$$10, 111011, 1111101111\ldots$$

Applying the reverse order procedure to this number pattern and arranging the given multiples of 9 in ascending order, yields a subsecondary sequence

$$9, 900, 90000\ldots$$

Now consider a nonstandard secondary number pattern

$$34, 1341, 113411, 11134111 \ldots$$

Suppose the natural order of terms of its subsecondary number pattern is

$$1, 3, 5 \ldots$$

Then the subsecondary number pattern is

$$34, 113411, 1111341111 \ldots$$

Applying the reverse order procedure to this number pattern and arranging the given multiples of 9 in ascending order, yields a subsecondary sequence

$$9, 900, 90000 \ldots$$

Since the two given subsecondary number patterns have the same natural order of terms and yield the same subsecondary sequence in the reverse order procedure, then Discovery 1 holds.

Definition 4: A pattern modifier of a given subsecondary number pattern is a nonzero digit that adjoins terms of the given subsecondary number pattern in a sequential manner to form terms of a new infinite number pattern.

Definition 5: Applying a pattern modifier to a given subsecondary number pattern means adjoining a pattern modifier and terms of the given subsecondary number pattern in a sequential manner.

Assumption: Adjoining a pattern modifier and terms of a given subsecondary number pattern in a sequential manner can be continued indefinitely.

The choice of a pattern modifier of a given subsecondary number pattern is arbitrary.

The following example gives meaning to the idea of applying a pattern modifier to a subsecondary number pattern.

EXAMPLE 10: Consider the standard secondary number pattern

$$10, 1101, 111011, 11110111 \ldots$$

Suppose 4 is applied 1 time to the given subsecondary number pattern. Then

$$1, 3, 5 \ldots$$

Then the subsecondary number pattern is

$$10, 111011, 1111101111 \ldots$$

Let 4 be the pattern modifier of the given subsecondary number pattern. Then new infinite number patterns arise as follows.

Suppose 4 is applied 1 time to the given subsecondary number pattern. Then this application produces a new infinite number pattern

$$4104, 41110114, 411111011114 \ldots$$

Suppose 4 is applied 2 times to the given subsecondary number pattern. Then this application produces a new infinite number pattern

$$441044, 4411101144, 44111110111144 \ldots$$

Suppose 4 is applied 3 times to the given subsecondary number pattern. Then this application produces a new infinite number pattern

$$444410444, 444111011444, 44411111011114444 \ldots$$

Applying 4 to the given subsecondary number pattern in the manner indicated can be done as many times as desired.

Discovery 2: The nth term of an infinite sequence of multiples of 9 which are generated with the reverse order procedure from a number pattern that is formed if a pattern modifier is applied y times to a given subsecondary number pattern, is 10^y as large as the nth term of a subsecondary sequence that is generated from the given subsecondary number pattern.

2. **The nth term of a subsecondary sequence that is generated from a subsecondary number pattern whose natural order of terms has first term 1 and common difference 2k, where k is a positive integer**

The nth term of the stated subsecondary sequence is based on the standard secondary number pattern

$$10, 1101, 111011, 11110111\ldots$$

Suppose the natural order of terms of its subsecondary number pattern has first term 1 and common difference 2. Then the natural order of terms of the subsecondary number pattern is

$$1, 3, 5\ldots$$

Hence the subsecondary number pattern is

$$10, 111011, 1111101111\ldots$$

Applying the reverse order procedure to this number pattern and arranging the given multiples of 9 in ascending order, yields a subsecondary sequence

$$9, 900, 90000\ldots$$

whose n^{th} term is

$$9(10^{2n-2})$$

for n = 1, 2 . . .

For the given secondary number pattern, suppose the natural order of terms of its subsecondary number pattern has first term 1 and common difference 4. Then the natural order of terms of the subsecondary number pattern is

$$1, 5, 9 \ldots$$

Hence the subsecondary number pattern is

$$10, 1111101111, 111111111011111111 \ldots$$

Applying the reverse order procedure to this number pattern and arranging the given multiples of 9 in ascending order, yields a subsecondary sequence

$$9, 90000, 900000000 \ldots$$

whose n^{th} term is

$$9(10^{4n-4})$$

for $n = 1, 2 \ldots$

We generalize from the two preceding n^{th} terms that the nth term of the stated subsecondary sequence is

$$9(10^{2nk-2k})$$

for $n = 1, 2 \ldots$, and $k = 1, 2 \ldots$

3. **The nth term of a subsecondary sequence that is generated from a subsecondary number pattern whose natural order of terms has first term 1 and common difference $2k + 1$, where k is a positive integer**

The nth term of the stated subsecondary sequence is based on the standard secondary number pattern

$$10, 1101, 111011, 11110111 \ldots$$

Suppose the natural order of terms of its subsecondary number pattern has first term 1 and common difference 3. Then the natural order of terms of the subsecondary number pattern is

$$1, 4, 7 \ldots$$

Hence the subsecondary number pattern is

$$10, 11110111, 11111110111111 \ldots$$

Applying the reverse order procedure to this number pattern and arranging the given multiples of 9 in ascending order, yields a subsecondary sequence

$$9, 9000, 9000000 \ldots$$

whose n^{th} term is

$$9(10^{3n-3})$$

for $n = 1, 2 \ldots$

For the given secondary number pattern, suppose the natural order of terms of its subsecondary number pattern has first term 1 and common difference 5. Then the natural order of terms of the subsecondary number pattern is

$$1, 6, 11 \ldots$$

Hence the subsecondary number pattern is

$$10, 111111011111, 1111111111101111111111 \ldots$$

Applying the reverse order procedure to this number pattern and arranging the given multiples of 9 in ascending order, yields a subsecondary sequence

$$9, 900000, 90000000000 \ldots$$

whose n^{th} term is

$$9(10^{5n-5})$$

for n = 1, 2 . . .

We generalize from the two preceding n^{th} terms that the nth term of the stated subsecondary sequence is

$$9(10^{(2k+1)n-(2k+1)})$$

for n = 1, 2 . . . , and k = 1, 2 . . .

4. **The nth term of a subsecondary sequence that is generated from a subsecondary number pattern whose natural order of terms has first term 2 and common difference 2k, where k is a positive integer**

The nth term of the stated subsecondary sequence is based on the standard secondary number pattern

$$10, 1101, 111011, 11110111 \ldots$$

Suppose the natural order of terms of its subsecondary number pattern has first term 2 and common difference 2. Then the natural order of terms of the subsecondary number pattern is

$$2, 4, 6 \ldots$$

Hence the subsecondary number pattern is

$$1101, 11110111, 111111011111 \ldots$$

Applying the reverse order procedure to this number pattern and arranging the given multiples of 9 in ascending order, yields a subsecondary sequence

$$90, 9000, 900000 \ldots$$

whose n^{th} term is

$$9(10^{2n-1})$$

for $n = 1, 2 \ldots$

For the given secondary number pattern, suppose the natural order of terms of its subsecondary number pattern has first term 2 and common difference 4. Then the natural order of terms of the subsecondary number pattern is

$$2, 6, 10 \ldots$$

Hence the subsecondary number pattern is

$$1101, 111111011111, 1111111110111111111 \ldots$$

Applying the reverse order procedure to this number pattern and arranging the given multiples of 9 in ascending order, yields a subsecondary sequence

$$90, 900000, 9000000000 \ldots$$

whose n^{th} term is

$$9(10^{4n-3})$$

for $n = 1, 2 \ldots$

We generalize from the two preceding n^{th} terms that the nth term of the stated subsecondary sequence is

$$9(10^{2kn-2k+1})$$

for $n = 1, 2 \ldots$, and $k = 1, 2 \ldots$

5. The nth term of a subsecondary sequence that is formed from a subsecondary number pattern whose natural order of terms has first

term 2 and common difference 2k + 1, where k is a positive integer

The nth term of the stated subsecondary sequence is based on the standard secondary number pattern

$$10, 1101, 111011, 11110111\ldots$$

Suppose the natural order of terms of its subsecondary number pattern has first term 2 and common difference 3. Then the natural order of terms of the subsecondary number pattern is

$$2, 5, 8\ldots$$

Hence the subsecondary number pattern is

$$1101, 1111101111, 1111111101111111\ldots$$

Applying the reverse order procedure to this number pattern and arranging the given multiples of 9 in ascending order, yields a subsecondary sequence

$$90, 90000, 90000000\ldots$$

whose n^{th} term is

$$9(10^{3n-2})$$

for n = 1, 2 . . .

For the given secondary number pattern, suppose the natural order of terms of its subsecondary number pattern has first term 2 and common difference 5. Then the natural order of terms of the subsecondary number pattern is

$$2, 7, 12\ldots$$

Hence the subsecondary number pattern is

$$1101,\ 11111110111111,\ 111111111111011111111111\ldots$$

Applying the reverse order procedure to this number pattern and arranging the given multiples of 9 in ascending order, yields a subsecondary sequence

$$90,\ 9000000,\ 900000000000\ldots$$

whose n^{th} term is

$$9(10^{5n-4})$$

for $n = 1, 2\ldots$, and $k = 1, 2\ldots$

We generalize from the two preceding n^{th} terms that the nth term of the stated subsecondary sequence is

$$9(10^{(2k+1)n-2k})$$

for $n = 1, 2\ldots$, and $k = 1, 2\ldots$

6. **The nth term of a subsecondary sequence that is generated from a subsecondary number pattern whose natural order of terms has first term 3 and common difference 3k, where k is a positive integer**

The nth term of the stated subsecondary sequence is based on the standard secondary number pattern

$$10,\ 1101,\ 111011,\ 11110111\ldots$$

Suppose the natural order of terms of its subsecondary number pattern has first term 3 and common difference 3. Then the natural order of terms of the subsecondary number pattern is

$$3, 6, 9\ldots$$

Hence the subsecondary number pattern is

$$111011, 111111011111, 111111111011111111\ldots$$

Applying the reverse order procedure to this number pattern and arranging the given multiples of 9 in ascending order, yields a subsecondary sequence

$$900, 900000, 900000000\ldots$$

whose n^{th} term is

$$9(10^{3n-1})$$

for $n = 1, 2\ldots$

For the given secondary number pattern, suppose the natural order of terms of its subsecondary number pattern has first term 3 and common difference 6. Then the natural order of terms of the subsecondary number pattern is

$$3, 9, 15\ldots$$

Hence the subsecondary number pattern is

$$111011, 111111111011111111, 111111111111111011111111111111\ldots$$

Applying the reverse order procedure to this number pattern and arranging the given multiples of 9 in ascending order, yields a subsecondary sequence

$$900, 900000000, 900000000000000\ldots$$

whose n^{th} term is

$$9(10^{6n-4})$$

for $n = 1, 2\ldots$

We generalize from the two preceding n^{th} terms that the nth term of the stated subsecondary sequence is

$$9(10^{3nk-3k+2})$$

for n = 1, 2 ... , and k = 1, 2 ...

7. **The nth term of an infinite sequence of multiples of 9 which are generated with the reverse order procedure from a number pattern that is formed if a pattern modifier is applied y times to a subsecondary number pattern whose natural order of terms has first term 1 and common difference 2k, where k is a positive integer**

The nth term of the stated infinite sequence of multiples of 9 is based on the standard secondary number pattern

$$10, 1101, 111011, 11110111 \ldots$$

Suppose the natural order of terms of its subsecondary number pattern has first term 1 and common difference 2. Then the natural order of terms of the subsecondary number pattern is

$$1, 3, 5 \ldots$$

Hence the subsecondary number pattern is

$$10, 111011, 1111101111 \ldots$$

Let 3 be the pattern modifier of this subsecondary number pattern. Then infinite sequences of multiples of 9 arise as follows.

Suppose 3 is not applied to the given subsecondary number pattern. Then the given subsecondary number pattern remains the same.

Applying the reverse order procedure to this number pattern and arranging the given multiples of 9 in ascending order, yields a subsecondary sequence

$$9, 900, 90000 \ldots$$

whose nth term is

$$9(10^{2n-2})$$

for n = 1, 2 . . .

Suppose 3 is applied 1 time to the given subsecondary number pattern. Then this application produces a new infinite number pattern

3103, 31110113, 311111011113 . . .

Applying the reverse order procedure to this number pattern and arranging the given multiples of 9 in ascending order, yields an infinite sequence

90, 9000, 900000 . . .

whose nth term is

$$9(10^{2n-1})$$

for n = 1, 2 . . .

Suppose 3 is applied 2 times to the given subsecondary number pattern. Then this application produces a new infinite number pattern

331033, 3311101133, 33111110111133 . . .

Applying the reverse order procedure to this number pattern and arranging the given multiples of 9 in ascending order, yields an infinite sequence

900, 90000, 9000000 . . .

whose nth term is

$$9(10^{2n})$$

for n = 1, 2 . . .

For the given secondary number pattern, suppose the natural order of terms of its subsecondary number pattern has first term 1 and common difference 4. Then the natural order of terms of the subsecondary number pattern is

$$1, 5, 9 \ldots$$

Hence the subsecondary number pattern is

$$10, 1111101111, 111111110111111111 \ldots$$

Let 3 be the pattern modifier of this subsecondary number pattern. Then infinite sequences of multiples of 9 arise as follows.

Suppose 3 is not applied to the given subsecondary number pattern. Then the given subsecondary number pattern remains the same.

Applying the reverse order procedure to this number pattern and arranging the given multiples of 9 in ascending order, yields a subsecondary sequence

$$9, 90000, 900000000 \ldots$$

whose n^{th} term is

$$9(10^{4n-4})$$

for n = 1, 2 . . .

Suppose 3 is applied 1 time to the given subsecondary number pattern. Then this application produces a new infinite number pattern

$$3103, 311111011113, 3111111110111111113 \ldots$$

Applying the reverse order procedure to this number pattern and arranging the given multiples of 9 in ascending order, yields an infinite sequence

$$90, 900000, 9000000000 \ldots$$

whose n^{th} term is

$$9(10^{4n-3})$$

for $n = 1, 2 \ldots$

Suppose 3 is applied 2 times to the given subsecondary number pattern. Then this application produces a new infinite number pattern

$$331033, 331111110111133, 3311111111101111111133 \ldots$$

Applying the reverse order procedure to this number pattern and arranging the given multiples of 9 in ascending order, yields an infinite sequence

$$900, 9000000, 90000000000 \ldots$$

whose n^{th} term is

$$9(10^{4n-2})$$

for $n = 1, 2 \ldots$

We generalize from the six preceding n^{th} terms that the nth term of the stated infinite sequence of multiples of 9 is

$$9(10^{2kn-2k+y})$$

which can be expressed as

$$10^y\{9(10^{2kn-2k})\}$$

for $n = 1, 2 \ldots$, $k = 1, 2 \ldots$, and $y = 0, 1, 2 \ldots$

It follows from the preceding n^{th} term that the nth term of an infinite sequence of multiples of 9 which are generated with the reverse order procedure from a number pattern that is formed if a pattern modifier is applied y times to a subsecondary number pattern whose natural order of terms has first term 1 and common difference 2k, is 10^y as large as the nth term of a subsecondary sequence that is generated from a subsecondary number pattern whose natural order of terms has first term 1 and common difference 2k, where k is a positive integer. Hence Discovery 2 of this Chapter holds.

8. **The nth term of an infinite sequence of multiples of 9 which are generated with the reverse order procedure from a number pattern that is formed if a pattern modifier is applied y times to a subsecondary number pattern whose natural order of terms has first term 2 and common difference 2k, where k is a positive integer**

The nth term of the stated infinite sequence of multiples of 9 is based on the standard secondary number pattern

$$10, 1101, 111011, 11110111 \ldots$$

Suppose the natural order of terms of its subsecondary number pattern has first term 2 and common difference 2. Then the natural order of terms of the subsecondary number pattern is

$$2, 4, 6 \ldots$$

Hence the subsecondary number pattern is

$$1101, 11110111, 111111011111 \ldots$$

Let 5 be the pattern modifier of this subsecondary number pattern. Then infinite sequences of multiples of 9 arise as follows.

Suppose 5 is not applied to the given subsecondary number pattern. Then the given subsecondary number pattern remains the same.

Applying the reverse order procedure to this number pattern and arranging the given multiples of 9 in ascending order, yields a subsecondary sequence

$$90, 9000, 900000 \ldots$$

whose n^{th} term is

$$9(10^{2n-1})$$

for $n = 1, 2 \ldots$

Suppose 5 is applied 1 time to the given subsecondary number pattern. Then this application produces a new infinite number pattern

$$511015, 5111101115, 51111110111115 \ldots$$

Applying the reverse order procedure to this number pattern and arranging the given multiples of 9 in ascending order, yields an infinite sequence

$$900, 90000, 9000000 \ldots$$

whose n^{th} term is

$$9(10^{2n})$$

for $n = 1, 2 \ldots$

Suppose 5 is applied 2 times to the given subsecondary number pattern. Then this application produces a new infinite number pattern

$$55110155, 551111011155, 5511111101111155 \ldots$$

Applying the reverse order procedure to this number pattern and arranging the given multiples of 9 in ascending order, yields an infinite sequence

$$900, 90000, 9000000 \ldots$$

whose nth term is

$$9(10^{2n+1})$$

for n = 1, 2 . . .

For the given secondary number pattern, suppose the natural order of terms of its subsecondary number pattern has first term 2 and common difference 4. Then the natural order of terms of the subsecondary number pattern is

$$2, 6, 10 \ldots$$

Hence the subsecondary number pattern is

$$1101, 111111011111, 1111111110111111111 \ldots$$

Let 5 be the pattern modifier of this subsecondary number pattern. Then infinite sequences of multiples of 9 arise as follows.

Suppose 5 is not applied to the given subsecondary number pattern. Then the given subsecondary number pattern remains the same.

Applying the reverse order procedure to this number pattern and arranging the given multiples of 9 in ascending order, yields a subsecondary sequence

$$90, 900000, 9000000000 \ldots$$

whose nth term is

$$9(10^{4n-3})$$

for n = 1, 2 . . .

Suppose 5 is applied 1 time to the given subsecondary number pattern. Then this application produces a new infinite number pattern

$$511015, 51111110111115, 5111111111101111111115\ldots$$

Applying the reverse order procedure to this number pattern and arranging the given multiples of 9 in ascending order, yields an infinite sequence

$$900, 9000000, 90000000000\ldots$$

whose n^{th} term is

$$9((10^{4n-2})$$

for $n = 1, 2\ldots$

Suppose 5 is applied 2 times to the given subsecondary number pattern. Then this application produces a new infinite number pattern

$$55110155, 5511111101111155, 551111111111011111111155\ldots$$

Applying the reverse order procedure to this number pattern and arranging the given multiples of 9 in ascending order, yields an infinite sequence

$$9000, 90000000, 900000000000\ldots$$

whose n^{th} term is

$$9(10^{4n-1})$$

for $n = 1, 2\ldots$

We generalize from the six preceding n^{th} terms that the nth term of the stated infinite sequence of multiples of 9 is

$$9(10^{2kn-2k+1+y})$$

which can be expressed as

$$10^y\{9(10^{2kn-2k+1})\}$$

for n = 1, 2 ..., k = 1, 2 ..., and y = 0, 1, 2 ...

It follows from the preceding n^{th} term that the nth term of an infinite sequence of multiples of 9 which are generated with the reverse order procedure from a number pattern that is formed if a pattern modifier is applied y times to a subsecondary number pattern whose natural order of terms has first term 2 and common difference 2k, is 10^y as large as the nth term of a subsecondary sequence that is generated from a subsecondary number pattern whose natural order of terms has first term 2 and common difference 2k, where k is a positive integer. Hence Discovery 2 of this Chapter holds.

9. **The nth term of an infinite sequence of multiples of 9 which are generated with the reverse order procedure from a number pattern that remains if the first y terms are dropped from a subsecondary number pattern whose natural order of terms has first term 1 and common difference 2k, where k is a positive integer**

The nth term of the stated infinite sequence of multiples of 9 is based on the standard secondary number pattern

10, 1101, 111011, 11110111 ...

Suppose the natural order of terms of its subsecondary number pattern has first term 1 and common difference 2. Then the natural order of terms of the subsecondary number pattern is

1, 3, 5 ...

Hence the subsecondary number pattern is

10, 111011, 1111101111 ...

Suppose no term is dropped from the given subsecondary number pattern.

Then the given subsecondary number pattern remains the same.

Applying the reverse order procedure to this number pattern and arranging the given multiples of 9 in ascending order, yields a subsecondary sequence

$$9, 900, 90000 \ldots$$

whose n^{th} term is

$$9(10^{2n-2})$$

for $n = 1, 2 \ldots$

Suppose the first term is dropped from the given subsecondary number pattern. Then the remaining number pattern is

$$111011, 1111101111, 11111110111111 \ldots$$

Applying the reverse order procedure to this number pattern and arranging the given multiples of 9 in ascending order, yields an infinite sequence

$$900, 90000, 9000000 \ldots$$

whose n^{th} term is

$$9(10^{2n})$$

for $n = 1, 2 \ldots$

Suppose the first 2 terms are dropped from the given subsecondary number pattern. Then the remaining number pattern is

$$1111101111, 11111110111111, 111111110111111111 \ldots$$

Applying the reverse order procedure to this number pattern and arranging the given multiples of 9 in ascending order, yields an infinite sequence

$$90000, 9000000, 900000000 \ldots$$

whose n^{th} term is

$$9(10^{2n+2})$$

for $n = 1, 2 \ldots$

For the given secondary number pattern, suppose the natural order of terms of its subsecondary number pattern has first term 1 and common difference 4. Then the natural order of terms of the subsecondary number pattern is

$$1, 5, 9 \ldots$$

Hence the subsecondary number pattern is

$$10, 1111101111, 111111110111111111 \ldots$$

Suppose no term is dropped from the given subsecondary number pattern. Then the given subsecondary number pattern remains the same.

Applying the reverse order procedure to this number pattern and arranging the given multiples of 9 in ascending order, yields a subsecondary sequence

$$9, 90000, 900000000 \ldots$$

whose n^{th} term is

$$9(10^{4n-4})$$

for $n = 1, 2 \ldots$

Suppose the first term is dropped from the given subsecondary number pattern. Then the remaining number pattern is

$$1111101111, 111111110111111111, 1111111111110111111111111 \ldots$$

Applying the reverse order procedure to this number pattern and arranging the given multiples of 9 in ascending order, yields an infinite sequence

$$90000, 900000000, 9000000000000 \ldots$$

whose n^{th} term is

$$9(10^{4n})$$

for $n = 1, 2 \ldots$

We generalize from the five preceding n^{th} terms that the nth term of the stated infinite sequence of multiples of 9 is

$$9(10^{2kn - 2k + 2yk})$$

for $n = 1, 2 \ldots, k = 1, 2 \ldots,$ and $y = 0, 1, 2 \ldots$

10. **The nth term of an infinite sequence of multiples of 9 which are generated with the reverse order procedure from a number pattern that remains if the first y terms are dropped from a subsecondary number pattern whose natural order of terms has first term 2 and common difference 2k, where k is a positive integer**

The nth term of the stated infinite sequence of multiples of 9 is based on the standard secondary number pattern

$$10, 1101, 111011, 11110111 \ldots$$

Suppose the natural order of terms of its subsecondary number pattern has first term 2 and common difference 2. Then the natural order of terms of the subsecondary number pattern is

$$2, 4, 6 \ldots$$

Hence the subsecondary number pattern is

1101, 11110111, 111111011111 . . .

Suppose no term is dropped from the given subsecondary number pattern. Then the given subsecondary number pattern remains the same.

Applying the reverse order procedure to this number pattern and arranging the given multiples of 9 in ascending order, yields a subsecondary sequence

$$90, 9000, 900000 \ldots$$

whose n^{th} term is

$$9(10^{2n-1})$$

for n = 1, 2 . . .

Suppose the first term is dropped from the given subsecondary number pattern. Then the remaining number pattern is

11110111, 111111011111, 1111111101111111 . . .

Applying the reverse order procedure to this number pattern and arranging the given multiples of 9 in ascending order, yields an infinite sequence

$$9000, 900000, 90000000 \ldots$$

whose n^{th} term is

$$9(10^{2n+1})$$

for n = 1, 2 . . .

Suppose the first 2 terms are dropped from the given subsecondary number pattern. Then the remaining number pattern is

111111011111, 1111111101111111, 11111111110111111111 . . .

Applying the reverse order procedure to this number pattern and arranging the given multiples of 9 in ascending order, yields an infinite sequence

$$900000, 90000000, 9000000000 \ldots$$

whose n^{th} term is

$$9(10^{2n+3})$$

for $n = 1, 2 \ldots$

For the given secondary number pattern, suppose the natural order of terms of its subsecondary number pattern has first term 2 and common difference 4. Then the natural order of terms of the subsecondary number pattern is

$$2, 6, 10 \ldots$$

Hence the subsecondary number pattern is

$$1101, 111111011111, 1111111110111111111 \ldots$$

Suppose no term is dropped from the given subsecondary number pattern. Then the given subsecondary number pattern remains the same.

Applying the reverse order procedure to this number pattern and arranging the given multiples of 9 in ascending order, yields a subsecondary sequence

$$90, 900000, 9000000000 \ldots$$

whose n^{th} term is

$$9(10^{4n-3})$$

for $n = 1, 2 \ldots$

Suppose the first term is dropped from the given subsecondary number pattern. Then the remaining number pattern is

$$111111011111, 11111111110111111111,$$
$$1111111111111101111111111111\ldots$$

Applying the reverse order procedure to this number pattern and arranging the given multiples of 9 in ascending order, yields an infinite sequence

$$900000, 9000000000, 90000000000000\ldots$$

whose n^{th} term is

$$9(10^{4n+1})$$

for $n = 1, 2 \ldots$

We generalize from the five preceding n^{th} terms that the nth term of the stated infinite sequence of multiples of 9 is

$$9(10^{2kn - 2k + 1 + 2yk})$$

for $n = 1, 2 \ldots, k = 1, 2 \ldots,$ and $y = 0, 1, 2 \ldots$

11. **The nth term of an infinite sequence of multiples of 9 which are generated with the reverse order procedure from a number pattern that remains if the first y terms are dropped from a subsecondary number pattern whose natural order of terms has first term 2 and common difference 2k + 1, where k is a positive integer**

The nth term of the stated infinite sequence of multiples of 9 is based on the standard secondary number pattern

$$10, 1101, 111011, 11110111\ldots$$

Suppose the natural order of terms of its subsecondary number pattern has first term 2 and common difference 3. Then the natural order of terms of the subsecondary number pattern is

$$2, 5, 8 \ldots$$

Hence the subsecondary number pattern is

$$1101, 1111101111, 1111111101111111 \ldots$$

Suppose no term is dropped from the given subsecondary number pattern. Then the given subsecondary number pattern remains the same.

Applying the reverse order procedure to this number pattern and arranging the given multiples of 9 in ascending order, yields a subsecondary sequence

$$90, 90000, 90000000 \ldots$$

whose n^{th} term is

$$9(10^{3n-2})$$

for $n = 1, 2 \ldots$

Suppose the first term is dropped from the given subsecondary number pattern. Then the remaining number pattern is

$$1111101111, 1111111101111111, 1111111111101111111111 \ldots$$

Applying the reverse order procedure to this number pattern and arranging the given multiples of 9 in ascending order, yields an infinite sequence

$$90000, 90000000, 90000000000 \ldots$$

whose n^{th} term is

$$9(10^{3n+1})$$

for n = 1, 2 . . .

Suppose the first 2 terms are dropped from the given subsecondary number pattern. Then the remaining number pattern is

$$1111111101111111, 111111111101111111111,$$
$$1111111111111101111111111111 \ldots$$

Applying the reverse order procedure to this number pattern and arranging the given multiples of 9 in ascending order, yields an infinite sequence

$$90000000, 90000000000, 90000000000000 \ldots$$

whose n^{th} term is

$$9(10^{3n+4})$$

for n = 1, 2 . . .

For the given secondary number pattern, suppose the natural order of terms of its subsecondary number pattern has first term 2 and common difference 5. Then the natural order of terms of the subsecondary number pattern is

$$2, 7, 12 \ldots$$

Hence the subsecondary number pattern is

$$1101, 11111110111111, 1111111111101111111111 \ldots$$

Suppose no term is dropped from the given subsecondary number pattern. Then the given subsecondary number pattern remains the same.

Applying the reverse order procedure to this number pattern and arranging the given multiples of 9 in ascending order, yields a subsecondary sequence

$$90, 9000000, 900000000000 \ldots$$

whose nth term is

$$9(10^{5n-4})$$

for n = 1, 2 . . .

Suppose the first term is dropped from the given subsecondary number pattern. Then the remaining number pattern is

11111110111111, 111111111111011111111111, 1111111111111111101 111111111111111 . . .

Applying the reverse order procedure to this number pattern and arranging the given multiples of 9 in ascending order, yields an infinite sequence

9000000, 900000000000, 90000000000000000 . . .

whose nth term is

$$9(10^{5n+1})$$

for n = 1, 2 . . .

We generalize from the five preceding nth terms that the nth term of the stated infinite sequence of multiples of 9 is

$$9(10^{(2k+1)n - 2k + (2k+1)y})$$

for n = 1, 2 . . . , k = 1, 2 . . . , and y = 0, 1, 2 . . .

RELATED BOOKS

Bluman, A.G. Pre-Algebra Demystified, Second Edition, The McGraw-Hill Companies, Inc., New York, NY, 2011.

Dantzig, T. Number the Language of Science, PLUME, Penguin Group (USA) Inc., 375 Hudson Street, New York, NY, 2007.

Fitzgerald, T.R. Math Dictionary For Kids, Prufrock Press Inc., P.O. Box 8813, Waco, TX, 2016.

Herzog, D.A. Teach Yourself VISUALLY Algebra, Wiley Publishing, Inc., 111 River Street, Hoboken, NJ, 2008.

Kelley, M.W. The Humongous Book of Basic Math & Pre-Algebra Problems, Penguin Group (USA) Inc., 375 Hudson Street, New York, NY, 2011.

CPSIA information can be obtained
at www.ICGtesting.com
Printed in the USA
BVHW050503080522
636345BV00008B/67